Living Legacies: A Multidisciplinary Perspective on Indian Knowledge Systems

Edited by

Dr Sigma G R
Shama Pillai

BLACK EAGLE BOOKS
Dublin, USA | Bhubaneswar, India

Black Eagle Books
USA address:
7464 Wisdom Lane
Dublin, OH 43016

India address:
E/312, Trident Galaxy, Kalinga Nagar,
Bhubaneswar-751003, Odisha, India

E-mail: info@blackeaglebooks.org
Website: www.blackeaglebooks.org

First International Edition Published by
Black Eagle Books, 2025

LIVING LEGACIES: A MULTIDISCIPLINARY PERSPECTIVE ON INDIAN KNOWLEDGE SYSTEMS
Edited by **Dr Sigma G R** | **Shama Pillai**

Cover & Interior Design: Ezy's Publication

ISBN- 978-1-64560-697-0 (Paperback)

Printed in the United States of America

CONTENTS

Introduction

Indian Knowledge Systems (IKS) are a large, dynamic, and changing system of thinking that has developed over many centuries based on self-experience, practice, and universal observation of nature and human beings. Crossing borders of disciplines like science, philosophy, literature, mathematics, agriculture, medicine, and linguistics, IKS is rather a living tradition than a historical legacy that continues to influence the intellectual and cultural life of the nation. This volume brings together 17 original and quality research papers that address several aspects of Indian Knowledge Systems from a range of disciplinary perspectives. The authors, representing as diverse an array of disciplines as English, Malayalam, Physics,Economis, Botany, History, Yoga and Mathematics, offer rich and nuancced understanding that sheds light on the richness and richness of IKS. Individually and collectively, their work bears witness to the fact that indigenous knowledge cannot be kept within the campuses of any single academic discipline—it is interdisciplinary and inclusive in nature.

At a time of fast pace technological growth and globalization-induced cultural change, there is greater urgency to turn back and reclaim the wisdom contained in indigenous systems. It is in that spirit that this book has been written, providing the reader with a multi-dimension view of IKS through academically grounded research. The chapters in this book are based on serious research, rigorous examination, and a profound respect for India's indigenous intellectual traditions. They are accepting a broad variety of themes—literary representations and historical reinterpretations, scientific research, and environmental understanding. Each paper not only speaks about the conceptual and practical aspects of IKS but also attempts to connect the gap between traditional wisdom and contemporary knowledge systems

Literary activity conducted by English and Malayalam departments explores the ways in which Indian knowledge traditions are grounded in texts, oral narratives, virtual world role, food culture and other discourses. They also examine the role played by literary traditions as a conduit for the transfer of IKS across generations and languages. Scientific articles in the field by scientists in Physics and Botany reveal how the ancient Indian processes of observation and experimentation were understated, methodical, and in most cases, ahead of their time. A scientist's investigation into ancient astronomy knowledge, metallurgy, nano technology unearths the accuracy and empiric basis of ancient Indian knowledge. Parallelly, a botanist's study of native plant taxonomy and herbal mythology brings to the fore the nature-oriented sensitivity and medical potency of ancient methodologies that continue to prevail in present-day health care and conservation practices. Sustainability and Environmental Economics in Indian Traditions is also discussed in this book .In the field of Mathematics, the study investigates the numeral systems, and algorithmic mind developed in Indian traditions—most of which led to the foundations of current methods of computation. These articles also show how ancient Indian mathematics was thoroughly interwoven with philosophical inquiry, and problem-solving in the real world. Contributions to the volume from history give a critical reading of Indian intellectual traditions, especially in the context of colonial disruption and postcolonial recovery. These articles talk about how colonial contact remodelled the course of indigenous knowledge systems.

Together, these 17 articles not only bring out the richness and diversity of Indian Knowledge Systems but also open new horizons for interdisciplinary research, curricula development, and policymaking. In considering IKS from both perspectives of tradition and innovation, this volume challenges educators, students, and scholars to rethink the very principles on which knowledge is based. Gathered in editing this volume, it is our intention to provide a deeper understanding of India's intellectual heritage and engage in an intellectual conversation respectful and renewing of indigenous culture. We trust that this book will stimulate more work, critical argument, and indeed a new awareness of the understanding which has underpinned Indian civilization through the ages.

We also thank all the contributors whose thoughtful research articles have enriched this book on Indigenous Knowledge Systems. We are also thankful to the supporters, well-wishers, and the publisher for their continued support and cooperation in this project. Their combined efforts have brought this book into existence. We are grateful to each one of them who directly or indirectly helped make this valuable work a reality.

Dr Sigma G R
Shama Pillai

Recasting Iron: Tamil Nadu's Archaeometallurgical Challenge to Global Chronologies

Dr. Chithra Lekha P
Senior Research Scientist,
Virginia Tech India, drlekhapc@gmail.com

Sanal K Mohanan
Chief Technical officer, Karotimam Innovations Pvt Ltd,
sanalkmohanan@gmail.com

Abstract

Recent archaeological and radiometric findings from Tamil Nadu have fundamentally reshaped the understanding of early iron metallurgy in India. While earlier narratives positioned the origin of Indian ironworking around 1200–1000 BCE and attributed its diffusion to West Asian influences, new evidence from sites such as Sivagalai, Adichanallur, and Mayiladumparai suggests the presence of an independent, indigenous ironworking tradition as early as 3345 BCE. This article frames Tamil Nadu's evidence within a multilinear model of archaeometallurgy, challenging the linear diffusionist paradigm and emphasizing the role of regional resources, social practices, and epistemologies. Drawing from stratigraphic data, Accelerator Mass Spectrometry (AMS), and rarely cited indigenous Siddha medical texts, the study highlights how local knowledge systems, refractory resource distribution, and ritual integration informed early technological innovation. Tamil Nadu's iron tradition thus not only extends the subcontinent's chronology of metallurgy but also calls for a decolonial, polycentric approach to global archaeometallurgical historiography.

1. Introduction: Rethinking the Origins of Metallurgy

The study of ancient metallurgy—once a pursuit dominated by linear chronologies and core-periphery diffusion models—has undergone a significant transformation over the last few decades. Archaeometallurgy, which originally sought to trace the singular "origin" of metal use in human societies, has evolved into a multidisciplinary and context-sensitive field. It now recognizes that technological innovation is rarely monolithic or unidirectional.

Rather, metallurgy developed independently across multiple geographies, shaped by local ecological settings, mineral availability, socio-political organization, and cultural values.

One of the most influential voices in early archaeological thought, V. Gordon Childe, proposed that civilization—including agriculture, urban life, and metallurgy—emerged in the Near East and gradually diffused outward to surrounding regions. While Childe's model was progressive for its time, emphasizing cultural transmission over racial determinism, it has since been critiqued for oversimplifying the complex and heterogeneous processes of technological innovation. Particularly in archaeometallurgy, there is growing consensus that innovations such as smelting and alloying were not necessarily borrowed from a single "cradle of civilization," but often emerged independently in response to local needs and resources.

This rethinking was formalized during the 2008 Society for American Archaeology (SAA) symposium on archaeometallurgy, which led to four paradigm-shifting conclusions. First, scholarship has shifted from seeking the origins of metallurgy to understanding the processes of **innovation, adaptation, and adoption**. Second, ancient technology must be understood within a **multi-craft framework**, wherein metallurgy co-evolved with ceramic production, lithic technologies, Siddha Medicine production and ritual practices. Third, long-standing assumptions linking metallurgy with elite control and social stratification must be critically examined—early metal use was not always tied to political power or centralized authority. Finally, a **holistic approach** to technological study is now considered essential, encompassing mining, smelting, toolmaking, and waste analysis, rather than focusing solely on finished artifacts.

This new direction in archaeometallurgy underscores the value of **interdisciplinary collaboration**, bringing together archaeology, mineralogy, anthropology, material science, geochemistry, and ethnography. It also calls for a more inclusive and decentralized narrative of metallurgical development, one that acknowledges the contributions of indigenous knowledge systems and lesser-studied regions such as South India.

2. **Global Context of Early Iron Metallurgy**

Traditional archaeological narratives have long outlined a neat, linear evolution of metallurgy—from the Stone Age through the Chalcolithic and Bronze Ages, culminating in the Iron Age. This copper ◉ bronze ◉ iron sequence, while broadly useful, does not fully capture the varied and complex metallurgical histories observed across different regions. Emerging archaeometallurgical evidence reveals that technological progress was often shaped by localized interactions between medicine synthesis, geology, societal needs, and innovation, rather than by a singular global trajectory.

The first metals to be widely used—such as copper, gold, and lead—were not selected for their abundance, but for their accessibility and ease of processing. These metals often occurred in native or easily smeltable forms and had low melting points, allowing early metallurgists to manipulate them with rudimentary furnaces. In contrast, iron, despite being more geochemically abundant, presented a technological challenge: it required higher temperatures and reducing conditions, delaying its widespread adoption.

Ancient societies did not classify ores by modern industrial standards. Rather, factors such as surface visibility, color, workability, and perceived utility shaped what was recognized as a usable ore. Malachite (copper) and hematite (iron) were favored for their distinct appearance and smelting compatibility. Depending on local geology, communities exploited igneous (magnetite, chromite), hydrothermal (sulfide), or sedimentary (laterite, bog iron) deposits. However, many early mining landscapes have been destroyed or obscured by later activity, complicating provenance studies.

Within this global context, **Anatolia (modern Turkey)** is widely regarded as the earliest center of systematic iron smelting. At **Kaman-Kalehöyük**, artifacts dating to 2200–2000 BCE demonstrate early steel production. Following the **Bronze Age collapse (~1200 BCE)**—which disrupted tin supplies—iron rapidly gained prominence across **West Asia**, replacing bronze as the primary material for tools and weapons.

From Anatolia, ironworking practices **diffused eastward into South Asia**, where sites like **Malhar** and **Atranjikhera** show iron artifacts by 1200–1000 BCE. In **Africa**, while some early iron dates are recorded, dominant theories support diffusion via **Egypt or trans-Saharan routes**, with ironworking well-established by 1000 BCE in the **Nok culture** of Nigeria.

In **East Asia**, iron metallurgy began around the **9th century BCE** in China, spreading to

Korea by the **4th century BCE** and reaching **Japan** during the **late Yayoi period (~300 BCE)**. In **Southeast Asia**, iron appeared between the **5th and 2nd centuries BCE**, introduced via **maritime trade with South Asia**. **Europe's Iron Age** commenced around **1100 BCE**, diffusing from the Balkans into Central and Northern Europe through the **Hallstatt** and **La Tène** cultures.

While this diffusion-based framework remains dominant, it is increasingly evident that **early metallurgy was not uniform nor unidirectional**. It was a **complex interplay of materials, techniques, cultural adaptation, and ingenuity**—a truth that continues to emerge from archaeological findings in previously underexplored regions, including recent discoveries from **Tamil Nadu in South India**, which challenge long-standing assumptions and invite deeper inquiry into local trajectories of technological evolution.

3. **Evolution of Iron in the Indian Subcontinent (Excluding Tamil Nadu)**

For much of the 20th century, the history of iron metallurgy in India was interpreted primarily through a diffusionist lens. According to this view, iron technology was introduced into the Indian subcontinent from West Asia around **1000–800 BCE**, following established trade and cultural routes, and gradually spread eastward and southward. This framework found archaeological support in the **Gangetic plains**, where iron artifacts— primarily in the form of tools, nails, and weapons—were discovered in association with the **Painted Grey Ware (PGW)** culture. Prominent sites such as **Atranjikhera**, **Raja Nala-kaTila**, and **Malhar** yielded substantial iron remains, leading scholars to designate this region as the core of early Indian ironworking.

In these northern sites, iron usage appears to coincide with increasing agricultural activity, urbanization, and socio-political complexity, particularly in the context of the later Vedic period. However, the presence of earlier copper-based Chalcolithic cultures in the region, such as the **Ochre Coloured Pottery (OCP)** and **Harappan remnants**, suggested a transitional phase in which copper preceded iron in both ritual and utilitarian roles. This sequence gave rise to a now-classical chronological model for India: Chalcolithic Iron Age Early Historic.

Moving further east and into central India, archaeological investigations in **Vidarbha (Maharashtra), Odisha, and Chhattisgarh** began to reveal a more complex and regionally varied picture. In these areas, associated with **Megalithic traditions**, iron artifacts often emerged in burial contexts, alongside large urns, stone circles, and memorial stones. Excavations uncovered **slag heaps, iron-smelting furnaces, tuyeres**, and a wide range of iron implements dating back to the **first millennium BCE**. These findings point to localized ironworking traditions that may have developed independently of the Gangetic sequence, possibly influenced more by tribal, pastoral, or agro-pastoral lifeways than by urban elite demand.

The iron technologies seen in these regions, though chronologically later than those in the Gangetic plains, were no less significant. The metallurgical knowledge reflected in the design of furnaces and the quality of finished tools indicates a sophisticated understanding of smelting and forging processes. The lack of large-scale copper-based industries in these areas further suggests that iron may not have been a secondary replacement but rather the **primary metallurgical innovation**. Such patterns challenge the idea of a uniform Chalcolithic-to-Iron trajectory across the subcontinent.

Collectively, the evidence from northern, central, and eastern India reflects a **plurality of ironworking traditions**, shaped by localized access to ore resources, varying social structures, and ecological conditions. These parallel metallurgical trajectories—some influenced by diffusion, others possibly indigenous—demonstrate the dynamic and decentralized character of early Indian metallurgy.

However, it is in **South India**, particularly in **Tamil Nadu**, that recent scientific excavations have pushed the antiquity of iron even further back—compelling scholars to reconsider long-held assumptions about the origin and spread of iron technology in the Indian subcontinent which lead to a strong support multilinear model..

4.   **Tamil Nadu's Iron Age: A Breakthrough in Antiquity**

The iron-rich landscape of Tamil Nadu has long held archaeological promise, but only recently has its global historical significance been scientifically established through stateof-the-art radiometric analysis. A landmark series of excavations, backed by meticulous Accelerator Mass Spectrometry (AMS) and Optically Stimulated

Luminescence (OSL) dating, has positioned Tamil Nadu as a key player in rewriting early iron technology narratives— not just within India, but across the ancient world.

Crucially, these datings were conducted at internationally recognized laboratories known for their precision and reliability. The AMS ^14C datings were performed at **Beta Analytic Radiocarbon Dating Laboratory**, Florida, USA, renowned for its High Probability Density (HPD) calibration methods. Complementing this, OSL datings were carried out at the **Birbal Sahani Institute of Palaeosciences (BSIP), Lucknow**, and the **Physical Research Laboratory (PRL), Ahmedabad**, both respected institutions in geochronological research.

Samples including paddy grains, ceramics, and charcoal collected from well-documented stratigraphic contexts—particularly from urn burials and habitation layers at sites like Sivagalai, Mayiladumparai, and Adichanallur—yielded calibrated dates as early as **midfourth millennium BCE (c. 3345–2953 BCE)**. These results challenge the conventional timeline of the Iron Age and suggest a potentially independent ironworking tradition in South India that predates established benchmarks by over a millennium.

The adoption of multiple laboratories and cross-validation through different dating techniques reflects the Tamil Nadu State Department of Archaeology's commitment to scientific rigor and international best practices in archaeometallurgical research.

The transformation began with coordinated exca-vations at key archaeological sites including **Sivagalai (Thoothukudi), Adichanallur (Tirunelveli), Mayiladump-arai (Krishnagiri), Kilnamandi (Tiruvannamalai),** and **Mangadu (Salem)**—stretching from the Western Ghats to the coastal plains of Tamil Nadu. These sites have yielded compelling evidence of early iron use in both habitation and burial contexts, fundamentally challenging established models of early iron technology.

What distinguishes these findings is not just their antiquity, but the **scientific rigor** and **global best practices** followed in their validation. The dating of iron artifacts, slag, and associated cultural materials was conducted through **Accelerator Mass Spectrometry (AMS)** and **Optically Stimulated Luminescence (OSL)** methods at **internationally recognized laboratories**. AMS radiocarbon dating was carried out at the **Beta Analytic Radiocarbon Dating Laboratory**, Florida, USA—renowned for its precision in archaeological chronologies. OSL dating was performed at India's premier geochronological research centers, including the **Birbal Sahani Institute of Palaeosciences (BSIP), Lucknow**, and the **Physical Research Laboratory (PRL), Ahmedabad**.

These high-precision datings have confirmed iron-working activities between **3345 BCE and 1510 BCE**, with the oldest evidence emerging from **Sivagalai**, firmly placing Tamil Nadu at the **forefront of global iron chronology**. The involvement of reputed scientific institutions underscores that this is not speculative archaeology, but a robust, data-driven breakthrough—establishing Tamil Nadu's Iron Age as a serious contribution to world archaeometallurgy.

At **Sivagalai**, archaeological teams uncovered urn burials containing iron fragments embedded in the cultural matrix of early farming communities. The associated dates, ranging between 3345 BCE and 2953 BCE, challenge the assumption that ironworking in India began only after the second millennium BCE. Similarly, **Adichanallur**, a site already renowned for its elaborate urn burials and bronze artifacts, yielded iron remnants dated to approximately **2517 BCE**. These were not isolated inclusions but appeared integrated into the daily and ritual fabric of early societies — suggesting functional and symbolic use of iron well before the supposed advent of the Iron Age in India.

Mayiladumparai, one of the earliest megalithic sites excavated in Krishnagiri district, pushed the boundaries further. Excavations there revealed iron slag, black-and-red ware pottery, and lateritic ore remnants associated with **2172 BCE**, confirming the site's metallurgical activity in the early 3rd millennium BCE. In **Kilnamandi**, a sarcophagus burial dated to **1692 BCE**contained well-preserved iron pieces, again indicating that metallurgy was deeply embedded in mortuary customs. Lastly, at **Mangadu**, an iron sword recovered from a disturbed burial layer yielded a calibrated date of **1510 BCE**, with accompanying signs of forging and bloomery activity.

Comparative Timeline of Early Iron Use

Region / Site	Earliest Iron Date (BCE)	Evidence Type
Sivagalai (Tamil Nadu, India)	3345–2953	Urn burials with iron, slag
Adichanallur (Tamil Nadu, India)	2517	Funerary urns, tools
Mayiladumparai (Tamil Nadu, India)	2172	Slag, ore, pottery
Kilnamandi (Tamil Nadu, India)	1692	Sarcophagus burial with iron
Mangadu (Tamil Nadu, India)	1510	Iron sword, bloomery

Anatolia (Kaman-Kalehöyük, Turkey)	1800–1500	Tools, slag
Mesopotamia (Tell Asmar, Iraq)	c. 2000 (isolated artifacts)	Iron tools and smelting residue (rare)
Nok Culture (Nigeria)	c. 1000	Terracotta figurines, smelting
China (Dinggong, Hebei)	c. 600	Iron knives, forge remains

*Note: Mesopotamian iron artifacts prior to 1500 BCE are rare and not indicative of widespread ironworking. Their presence likely represents early experimentation or tradebased acquisition rather than local smelting traditions.

What makes these findings exceptional is not merely their antiquity, but the technological sophistication they represent. Across these sites, archaeologists documented **smelting furnaces, tuyere pipes, slag deposits, crucible fragments**, and in some cases, evidence of **high-carbon content in the iron**, pointing toward an early understanding of carburization—essential for producing **wootz-like steel**. The widespread presence of **hematite, magnetite, and laterite ores** in Tamil Nadu's terrain provided easy access to raw material, while the development of controlled high-temperature furnaces suggests an advanced command over pyrotechnology.

Another critical aspect of Tamil Nadu's Iron Age is its **integration with ritual and burial practices**. The close association between iron implements and funerary customs— especially in urn and sarcophagus burials— indicates that iron had both utilitarian and cultural value. Rather than being introduced abruptly as a new technological phase, iron appears to have emerged organically within the social and spiritual lives of these early Tamil communities. This nuanced role of iron, not just as a tool or weapon but as a **marker of identity and belief**, reveals a society that had internalized the metallurgical craft at multiple levels.

These findings call into question long-standing assumptions about the timing and trajectory of iron technology. They provide empirical evidence against the diffusionist model that assumes iron arrived in India from the West around 1000 BCE. Instead, the Tamil Nadu sites demonstrate that **iron technology could have evolved independently in the Indian peninsula**, likely through indigenous experimentation while making pottery and Siddha medicine synthesis, resourcefulness, and social necessity.

Moreover, this regional breakthrough repositions South India on the global map of early metallurgy. When viewed alongside contemporaneous ironworking traditions in Anatolia, China, and Sub-Saharan Africa, the Tamil Nadu evidence suggests a **polycentric model of technological innovation**, where multiple civilizations arrived at iron independently, shaped by their distinct ecological and cultural contexts. It urges scholars to abandon Eurocentric periodizations and embrace a more **decentralized, pluralistic understanding** of human technological history and evolution.

In conclusion, the Tamil Nadu excavations are more than a revision of dates; they are a revision of thought. By combining rigorous fieldwork with advanced dating techniques and a culturally sensitive interpretive framework, these discoveries offer a compelling case for **reimagining the global history of iron**. Tamil Nadu's Iron Age is no longer a peripheral chapter in South Asian history—it is a **cornerstone in the evolving narrative of early metallurgy**.

5. Scientific Reviews and Global Reactions

The recent publication of Tamil Nadu's *Antiquity of Iron* report has reverberated across the global archaeometallurgical community. For the first time, a state-level archaeological initiative has presented robust radiometric data that not only pushes back the timeline of iron metallurgy in the Indian subcontinent but challenges the broader global consensus on the origin and spread of early iron technology. The reactions from international experts have ranged from cautious acknowledgment to intrigued skepticism—marking a significant shift in how regional research in South Asia is being received globally.

5.1 Endorsement with Caution: Professor David Killick's Review

Among the most notable reviews is that of Professor David Killick, a leading archaeometallurgist from the University of Arizona. Killick, widely respected for his work on ancient smelting technologies and ore processing, reviewed the Tamil Nadu report in detail and offered a scientifically balanced, albeit cautiously framed, perspective.

Killick affirms several critical aspects of the report:

- He accepts "without doubt" that iron was in use in Tamil Nadu by 1300 BCE, based on AMS dates from millet and rice within funerary urns.
- He considers the dates from Kilnamandi and Mayiladumparai (1850–1600 BCE) to be valid and methodologically sound.
- Most strikingly, he does not dismiss the earliest charcoal dates (~2500 BCE and earlier) from sites like Sivagalai and Adichanallur, noting that the probability of three separate samples across varied contexts all being impacted by the "old wood" effect is statistically unlikely.

Killick also praises the rigor of archaeological fieldwork and the team's meticulous documentation, indicating a high level of confidence in the procedural quality of the Tamil Nadu excavations.

5.2 Critical Evaluation: A Case of Interpretive Bias

While Killick's endorsement of the radiometric dates marks an important moment of scholarly validation, his interpretive suggestions reflect certain enduring biases that persist in global archaeometallurgy. Specifically, Killick proposes that Harappan metallurgists may have migrated south—hypothetically bringing copper-based metallurgical knowledge with them—and that in the absence of copper in Tamil Nadu, this expertise was redirected toward ironworking.

This line of reasoning, while attempting to explain technological discontinuity, is problematic for several reasons:

Resource Misconception

Killick's assertion overlooks Tamil Nadu's well-documented wealth of iron ore resources, including hematite, magnetite, and lateritic deposits, many of which are surface-exposed and require no deep mining. The region may lack copper in abundance, but its metallogenic landscape is ideally suited for iron smelting—making it plausible that communities developed metallurgical strategies directly suited to available resources, rather than relying on imported knowledge or materials.

In particular, **Salem and Tiruvannamalai districts** are rich in **banded magnetite formations**, a form of iron ore highly suitable for early smelting. In **Kanjamalai (Salem district)**, three distinct bands of magnetite contain inferred reserves of approximately **50– 60 million tonnes**, with iron content ranging from **33–36%**. Nearby, **Godumalai** features tightly folded magnetite bands with an additional **60–70 million tonnes** of inferred reserves. In **Tiruvannamalai district**, the hills of **Kavuthimalai and Vediappanmalai** host **banded magnetite quartzite with hematite**, distributed across three separate geological basins, accounting for around **60 million tonnes** of reserves.

Iron ore has also been reported in other districts such as **Dharmapuri, Erode, Nilgiris, Tiruchirapalli, and Villupuram**, according to data from the **ENVIS Centre Tamil Nadu**. This widespread and abundant presence of iron-rich geology makes it highly unlikely that Tamil metallurgists were dependent on knowledge transfer from copperbased traditions elsewhere. Instead, it reinforces the case for **locally-driven innovation**, adapted to the terrain and materials native to the region.

In addition to its iron wealth, **Tamil Nadu also possesses abundant deposits of refractory-grade materials** such as **high-quality clay, quartz, and graphite**—further reinforcing the idea that early metallurgists had access not only to ores but also to the **essential materials for constructing durable, heat-resistant furnaces** and crucibles. These materials form the foundation of **refractory technologies**, enabling hightemperature metallurgical processes vital to smelting, alloying, and metal purification.

- Graphite is found in significant quantities in **Sivaganga** and **Madurai** districts, with the **Kurunjakulam Graphite Block** in **Tenkasi district** recognized as a major source of high-quality crystalline graphite. Notably, **Sivagangai's TAMIN refinery** currently produces graphite of up to **96% purity** using **environmentally friendly water-based and alkaline processing methods**, eliminating the need for high-temperature treatments traditionally used in refining. The geological advantage of this region is further underscored by the fact that **graphite deposits are available just 3 feet below the surface**, making extraction highly accessible and cost-effective. Renowned for its ability to withstand temperatures exceeding **2700°C in inert environments**, graphite remains a crucial material in the construction of **crucibles, electrodes, and high-temperature reactors**, valued for its **exceptional thermal resistance and chemical stability**. The easy availability of such refractory-grade graphite in Tamil Nadu reinforces the likelihood that ancient metallurgists in the region had consistent access to superior refractory materials—essential for constructing durable smelting installations and mastering high-temperature pyrotechnology.

- **Fireclay**, used in traditional furnace construction, occurs in **Tiruvannamalai district** and supports **pottery-making traditions in Manamadurai (Sivaganga district)**. Fireclay refractories are capable of enduring **temperatures exceeding 1700°C**, making them ideal for **lining metallurgical kilns and bloomery furnaces**.

- **Quartz**, a high-melting-point silica mineral with low thermal expansion, is mined in **Kendenahalli and Ramakondahalli villages (Pennagaram Taluk)** and in the **Kattumunnur Mines of Karur district**. Used in **smelting installations and ceramic molds**, quartz provides **dimensional stability under heat**, a key trait in primitive and advanced furnace technologies alike.

Refractory and Metallurgical Resource Distribution in Tamil Nadu

District / Region	Resource Type	Estimated Reserves / Notes
Salem – Kanja-malai	Iron ore (banded magnetite)	50–60 million tonnes, 33–36% Fe
Salem – Goduma-lai	Iron ore (banded magnetite)	60–70 million tonnes, folded bands
Tiruvannamalai – Kilnamandi	Iron ore, Fireclay	Iron from sarcophagus; Fireclay tradition
Tiruvannamalai – Kavuthimalai, Vediappanmalai	Iron ore (banded magnetite quartzite, hematite)	Approx. 60 million tonnes in 3 basins
Sivaganga – Manamadurai	Graphite, Fireclay, Quartz (used in pottery & kilns)	Supports pottery and refractory traditions
Madurai &Tenkasi – Kurunjakulam Block	Graphite (crystalline, high-grade)	Used in crucibles, thermal components
Pennagaram Taluk – Kendenahalli& Ramakondahalli	Quartz and Feldspar (refractory)	Used in smelting installations, ceramic molds
Karur – Kattumunnur Mines	Quartz (high purity)	Refractory-grade quartz for metallurgy

This rich availability of **natural refractory materials** in Tamil Nadu provides further evidence that local ironworking communities likely developed **indigenous hightemperature technologies**, independent of external knowledge systems. Such resources would have enabled the construction of sophisticated smelting installations capable of producing not just bloom iron but potentially **high-carbon steels**—as suggested by archaeological findings. The presence of high-quality graphite might have accelerated the reduction of hematite and magnetite while working on pottery or Siddha medicine products like *bhasmam* or *sindhooram*.

Under appreciation of Indigenous Knowledge Systems

Killick's analysis also fails to account for the rich indigenous knowledge systems that have long informed medicinal, botanical, and artisanal practices in South India. The use of iron in traditional Siddha medicine, early surgical instruments, and agricultural tools suggests a deep and practical understanding of metal properties long before standardized scripts or centralized metallurgical guilds emerged. These cultural-technological continuities imply that the knowledge of materials, fire, and transformation was rooted in local epistemologies, not solely in Northern or external traditions.

The limitations of contemporary archaeometallurgical interpretations—such as those subtly embedded in Killick's review—highlight a critical gap in the scientific community's engagement with **indigenous knowledge systems**, particularly those preserved in traditional medical and alchemical texts. Written documentation from as early as **3500–2000 BCE** is exceedingly rare across civilizations, including the Indian subcontinent. During this period, **knowledge transmission was predominantly oral**, with specialized skills and techniques passed down through generations within practitioner communities. This oral tradition played a pivotal role in safeguarding complex empirical knowledge, which was only later codified into written form, especially within systems like **Siddha medicine**.

An example of this continuity is found in the **Siddha treatise Yacob**, a Tamil text dating back centuries, which contains a **poetic yet technically precise description** of furnace construction for high-temperature operations— specifically designed for processing **sulphur, orpiment, and metals** associated with Siddha alchemy. Such texts reflect a long-standing and locally developed understanding of **pyrotechnology, material science, and chemical transformation**, deeply integrated into medicinal practices and metallurgy. The endurance of these knowledge systems—despite the absence of early written records— reinforces the importance of **revisiting traditional medical literature and oral histories** as valuable sources for reconstructing ancient technological expertise.

One striking verse from *Yacob* describes the creation of a furnace made of a mixture of **fine clay, powdered brick, calcinated shell powder**, and treated with **herbal decoctions** such as *kadukkaikashayam*. This furnace, once cured and sealed, is said to withstand repeated use and even **lightning strikes**—a metaphorical way of describing its thermal resilience and structural strength. The poetic account aligns remarkably well with modern definitions of **refractory materials**, pointing to a **technological continuity** in Tamil Nadu that modern science is only beginning to appreciate.

Yet, such sources are rarely cited or examined in mainstream archaeometallurgical discourse. Their absence not only reinforces the epistemological divide between Western scientific paradigms and South Asian indigenous systems but also leads to **mischaracterizations of local innovations** as "unexpected" or "unexplained." Notably this opens a window to the need of much systematic study of Siddha system and buried knowledge.

To move forward, scholars must expand their source base to include **regional textual traditions** like *Siddha, Ayurveda, Rasashastra,* and *Bhuta Vijnana*—texts that often encode empirical, experimental practices in metaphorical or symbolic language. These writings provide valuable clues to **materials science, thermodynamics, and metallurgical techniques** developed in ancient India, and may help fill the interpretive gaps currently bridged with speculative hypotheses like Harappan migration.

Reinforcement of Core–Periphery Thinking

The suggestion of Harappan migration inadvertently reinstates a core–periphery model in which innovation is presumed to originate from the northwest and disseminate southward.

This perspective echoes older diffusionist models that recent scholarship—including Killick's own advocacy for regional archaeometallurgical histories—has tried to move beyond. Ironically, while acknowledging the validity of Tamil Nadu's dates, Killick defaults to an explanatory framework that contradicts the polycentric evolution of technology he otherwise supports.

5.3   Toward a Decolonial Archaeometallurgy

The reception of Tamil Nadu's findings, including Killick's review, illustrates a broader need to decolonize global archaeological thought. Regional centers of innovation—especially in South Asia, Africa, and Southeast Asia—have long been subjected to interpretive frameworks developed in the West. Tamil Nadu's data offer not just a chronological correction but an epistemic challenge: they compel scholars to acknowledge that indigenous technological trajectories may not align with expected evolutionary sequences and must be understood on their own terms.

Rather than explaining away these findings through

Northern antecedents or Harappan legacies, it is imperative to recognize Tamil Nadu as a primary center of metallurgical innovation, where the leap into ironworking was facilitated not by external inputs but by internal creativity, geological abundance of minerals on the surface, and culturally embedded scientific practice.

Tamil Nadu's *Antiquity of Iron* report thus serves not just as a breakthrough in field archaeology, but also as a **call to revisit the literary and medicinal corpus of India**, which may hold critical insights into one of humanity's most transformative technologies: **the mastery of fire and metal**.

Conclusion: Recasting the Origins of Iron through a Multilinear Lens

The findings from Tamil Nadu compel a fundamental revision of the archaeological understanding of iron metallurgy in India and beyond. Far from being a late adopter of metallurgical technologies diffused from the northwest, South India—through its sites like Sivagalai, Adichanallur, and Mayiladumparai—demonstrates an indigenous and contextspecific trajectory of iron innovation that dates back to the **mid-4th millennium BCE**.

This evidence underscores the limitations of **linear, diffusionist models** that have long dominated global archaeometallurgical narratives. Instead, the South Indian case exemplifies a **multilinear model**, where technological innovation arises independently across multiple geographies, driven by local material resources, socio-cultural dynamics, and epistemic traditions. The integration of iron into ritual life, mortuary practices, and subsistence economies in Tamil Nadu illustrates that metallurgy was not merely a technological feat, but also a **deeply cultural and symbolic process**.

Moreover, the region's rich reserves of iron ore, coupled with natural deposits of **refractory materials** like **graphite, quartz, and fireclay**, reveal an ecosystem primed for early hightemperature experimentation. These natural affordances, combined with **indigenous knowledge systems** such as Siddha alchemy—evident in texts like *Yacob*—point to a sophisticated understanding of heat, material transformation, and metallurgical control that predates formal state structures or urbanism.

The Tamil Nadu excavations thus not only push back the chronology of iron in the Indian subcontinent but also provide a **template for decolonizing archaeometallurgy**. By validating regional centers of innovation and revisiting indigenous textual traditions, this research invites a more plural, nuanced, and equitable understanding of technological history—one in which Tamil Nadu stands not at the margins, but at the **very heart of early global ironmaking**.

References

- **Childe, V. Gordon.***Man Makes Himself.* London: Watts & Co., 1936.
- **Killick, David J.**
- "Review of 'Antiquity of Iron' (Tamil Nadu State Archaeology Report, 2025)." *Frontline*, March 2025.
- "Invention and Innovation in African Iron-smelting Technologies." *Cambridge Archaeological Journal* 25, no. 1 (2015): 307–319.
- **Tamil Nadu State Department of Archaeology.** *Antiquity of Iron: Recent Radiometric Dates from Tamil Nadu.* Government of Tamil Nadu, 2025.
- **Rajagopal, R.** "Ancient Iron and Steel Technology in

South India." *Indian Journal of History of Science* 38, no. 4 (2003): 345–365.

- **Liu, Yong.** "The Development of Early Iron Technology in Ancient China." In *Archaeometallurgy in Global Perspective*, edited by Benjamin W. Roberts and Christopher P. Thornton. Springer, 2014.
- **Okafor, Emmanuel E.** "Nok Culture and Early Iron Working in Nigeria." *Journal of African Archaeology* 2, no. 1 (2004): 111–135.
- **Hegde, K. T. M.** "Iron Technology in Ancient India." In *History of Science, Philosophy and Culture in Indian Civilization*, Vol. I, Part 2. Centre for Studies in Civilizations, 1999.
- **Boivin, Nicole.***Material Cultures, Material Minds: The Impact of Things on Human Thought, Society and Evolution.* Cambridge University Press, 2008.
- **ENVIS Centre Tamil Nadu.** "Mineral Resources." Ministry of Environment, Forests and Climate Change. http://tnenvis.nic.in
- **Yacob Siddha Manuscript.** Published by Thamarai Noolagam, Chennai, 2025 (Tamil).
- **Ratnagar, Shereen.***Makers and Shapers: Early Indian Technology in the Home, Village and Urban Workshop.* New Delhi: Tulika Books, 2007.
- **Tylecote, R. F.** "Iron in the Old World." In *Ancient Iron Production and Processing in the Old World*, edited by B. W. Roberts and C. P. Thornton. Springer, 2014.
- **Crew, P. and Salter, C.** "The Little Bowl That Could! Experimental Iron Smelting in a Bowl Furnace." *Historical Metallurgy Society*, 2015.

Going Back to the Roots: Ayurveda and Traditional Food Patterns as Panacea forLifestyle Diseases in Kerala

Dr. Raj Sree M S
Associate Professor of English, All Saints' College

Abstract

This study investigates the transformation of Kerala's food culture through the lens of cultural food theories, emphasizing the resurgence of traditional cooking patterns and the Ayurvedic approach to food. With increasing cases of lifestyle diseases and dietary disorders linked to globalization and culinary Westernization, this study argues that a return to Kerala's indigenous food systems offers a culturally grounded and health-sustaining alternative. Through frameworks like Gastro-Nationalism and Culinary Cosmopolitanism, this study situates Ayurveda and traditional practices at the heart of a contemporary food revival that challenges the homogenizing pressures of global food trends.

Keywords: Ayurveda, Gastro-Nationalism, Culinary Cosmopolitanism, Food, Tradition.

Introduction

Basic necessities such as food, shelter, and clothing are essential for human survival. However, beyond their functional roles, these necessities have evolved into defining elements of culture, shaping societies, identities, and traditions across the world. Across centuries, crossing borders and boundaries, food, shelter, and clothing transcended their basic survival purposes to become integral parts of cultural frameworks, reflecting historical, environmental, and social influences. They become cultural markers, representing historical evolution and societal priorities. Globalization has led to a fusion of styles, where the traditional and the contemporary blend to create new cultural expressions towards creating identity, social status, and cultural heritage.

Food, shelter, and clothing are more than mere necessities; These elements shape human interactions, traditions, and identities, illustrating the dynamic relationship between survival needs and cultural evolution. Food is not just for sustenance; it is an expression of cultural identity, heritage, and social bonds. Different cuisines emerge based on geographical conditions, resource availability, and historical exchanges. For example, Mediterranean cuisine heavily features olive oil, seafood, and grains due to the region's climate and agricultural practices. Similarly, spices play a significant role in Indian cuisine, a reflection of both indigenous traditions and trade influences. Use of Coconut and coconut oil in Kerala cuisine is yet another example. Food also plays a central role in rituals, celebrations, and religious practices.

Hypothesis

In response to the dietary and health disruptions caused by globalized eating habits, Kerala is experiencing a cultural and nutritional realignment through the revival of Ayurvedic principles and traditional food practices as a holistic solution to lifestyle diseases.

Research Objectives

- To analyze Kerala's shifting food practices through cultural, health, and environmental frameworks.
- To evaluate the influence of globalization and processed foods on dietary health and lifestyle diseases in Kerala.
- To explore the resurgence of Ayurvedic and traditional food practices as a cultural and therapeutic response.
- To assess how food becomes a site of identity, memory, and resistance through the lens of Gastro-Nationalism and Culinary Cosmopolitanism.

Significance of the Study

As Kerala grapples with rising instances of obesity, diabetes, hypertension, and digestive disorders—ailments often traced to the nutrition transition—this study provides critical insights into how traditional foodways offer a sustainable and culturally resonant response. Ayurveda's integrative dietary approach, rooted in bio-individuality and seasonal adaptation, positions it as a powerful countermeasure to the health consequences of modern food systems. This study has contemporary relevance taking into consideration Kerala's changing food patterns. By contextualizing these practices within cultural theory, this study sheds light on how a return to the roots is both a political and practical imperative for wellness, food sovereignty, and cultural continuity.

Research Methodology

This qualitative study is based on ethnographic observation, in-depth interviews with Ayurvedic doctors, home cooks, and food activists, and analysis of social media trends and wellness literature. Archival materials and government health reports supplement the primary data, while content analysis of food blogs and wellness platforms provides insight into the contemporary discourse around traditional diets and Ayurveda in Kerala.

Theoretical Framework

Classical theorists such as Lévi-Strauss, Mary Douglas, and Bourdieu frame food as a cultural code that reflects and structures social life. Lévi-Strauss's culinary triangle reveals how cooking is a mode of cultural classification, while Douglas links food rituals to boundaries and social hierarchies. Claude Lévi-Strauss and Mary Douglas emphasize food's symbolic function. Traditional meals like Sadhya signify ritual purity and communal identity, whereas modern fast foods embody convenience and status.

Pierre Bourdieu's theory of taste links food preferences to class distinctions. In Kerala, food choices are stratified by income and cultural capital, distinguishing organic and Ayurvedic diets from fast food consumption. Bourdieu's concept of distinction illustrates how taste operates as a marker of class.

Claude Fischler's notion of the "omnivore's paradox" and his theory on social construction of food capture the tension between dietary variety and anxiety in the age of globalization. Food is a medium of memory and identity. Dishes like Puttu-Kadala are cultural anchors, whereas foods like pizza and fried chicken symbolize modernity.

Nutrition Transition Theory, as developed by Barry Popkin, elucidates the health challenges emerging from dietary shifts toward ultra-processed foods and sedentary lifestyles. Economic development has caused a shift from traditional rice-based meals to processed, highcalorie foods, resulting in lifestyle diseases.

Meanwhile, Appadurai's theories of Gastro-Nationalism and Culinary Cosmopolitanism help decode how nations assert identity through food, even while adapting to global influences. Gastro-Nationalism emphasizes food as a marker of national identity, resisting globalization. Culinary Cosmopolitanism emphasizes openness to global cuisines and culinary fusion. These theoretical frames set the stage for interpreting Kerala's current food dynamics, where Ayurveda and ancestral eating patterns re-emerge as tools for health restoration and cultural resilience.

Discussion

Kerala's culinary traditions are inseparable from its agrarian calendar, climatic diversity, and spiritual philosophies, particularly Ayurveda. Historically, food in Kerala was not merely sustenance but medicine and ritual, tuned to the cycles of nature and individual constitution (prakriti). Meals like Kanji-Payar, Njavara rice gruel, or Koovapodi were functional foods designed for detoxification, digestion, and immunity.

The waves of culinary transformation in Kerala include the traditional agrarianAyurvedic food system, colonial and trade influences, Gulf migration and Middle Eastern culinary integration, and the current phase of globalization marked by both cosmopolitan tastes and a revival of traditional and Ayurvedic practices. There are four major waves of change in food pattern in Kerala.

First Wave: Evolution of Porotta and Biriyani

The combination of porotta and beef has a special place in the hearts of Keralites. This dish transcends class, religion, and social status, making it not just a meal but a cultural phenomenon. The layered, flaky porotta (or parotta) is believed to have South Indian and Sri Lankan origins, influenced by Malabar cuisine and Tamil street food (where it is known as "parotta"). Porotta gained popularity due to its affordability, easy availability in roadside thattukadas (street food stalls), and ability to pair well with spicy curries. The layered texture and crisp-yet-soft bite made it an instant favorite. Porotta and beef as socio-cultural unifiers across class and religion. Porotta and Beef became ore than just a meal. The dish is a staple among daily wage laborers, auto drivers, and college students because it is cheap, filling, and widely available at local eateries. Kerala's vibrant street food scene is dominated by small thattukadas that serve porotta with spicy beef curry or beef fry. These stalls often stay open late at night, making them social hubs. Whether it's a late-night snack, a celebration meal, or a casual dinner, porotta and beef are comfort foods for many Keralites. The dish is equally popular at elite restaurants and roadside shacks. The dish brings together different communities, economic classes, and generations, making it an enduring icon of Kerala's culinary identity. Whether at a bustling thattukada or a high-end restaurant, the love for porotta and beef remains strong, proving that food can be more than sustenance—it can be a symbol of culture, identity, and unity. Along side we have our all time favourite Biriyani. Biryani holds a special place in Kerala's culture and cuisine, blending Mughlai, Persian, and South Indian influences with a unique local touch. It is not just a dish but a symbol of celebration, enjoyed during festivals, weddings, and special occasions across all communities.

Second Wave: Rise of Chinese Cuisine

By the late 20th century, Chinese cuisine started gaining popularity in Kerala, primarily due to urbanization, the influence of Indian-Chinese restaurants, and the appeal of fast food. Dishes like fried rice, noodles, and chili chicken became favourites among younger generations.

Many local restaurants started adapting these dishes with Indian flavours, leading to what is commonly known as "Indo-Chinese" cuisine in Kerala. The affordability, ease of preparation, and acceptance of Chinese flavours made it a dominant food trend in the early 2000s.

Third Wave: Thai, Continental and North Indian Influences

Thai food has found a niche among urban food lovers due to its balance of flavours, aromatic ingredients, and unique cooking techniques. The increasing popularity of Thai food in Kerala is due to the natural similarities between the two cuisines: use of Coconut, Rice, spices and Seafood. Nevertheless faced many challenges like ingredient availability. Certain Thai ingredients like galangal, kaffir lime leaves, and Thai basil are not easily available in Kerala, leading to substitutions. Moreover, Kerala's spice profile is different from Thai cuisine, requiring chefs to balanceflavors to suit local tastes. Authentic Thai food in Kerala is mostly limited to highend restaurants due to the cost of imported ingredients

North Indian dishes have seamlessly integrated into Kerala's food culture, enriching its already diverse cuisine. Dishes like paneer butter masala, naan, and chole bhature, pani puri, pav bhaji, samosa etc are all time favourite of people in Kerala.

Fast food giants like KFC and McDonald's have significantly influenced Kerala's food culture changing eating habits, lifestyles, and even inspiring local adaptations. KFC,

McDonald's, Subway and Dominos have made Western fast food a part of everyday dining, especially among the youth. They became a preferred option for quick bites, casual meet-ups, and family outings, shifting some traditional food habits.

Fourth Wave: Taking over of Mandi

But Mandi became a one word for porotta, Sadhya, biriyani and Chineese a kind of better replacement on account of quantity and affordability. In the last decade, Mandi rice, an Arabic origin dish, has become a major trend in Kerala, particularly in urban areas and among younger food enthusiasts.

Several factors contributed to this shift. Primarily the Gulf influence. A large number of Keralites work in Gulf countries and have developed a taste for Middle Eastern cuisine. They recreated dishes like Mandi, Shawarma, and Kuboos. With the rise of food vlogging and social media, Mandi became a trending dish, with new restaurants specializing in Arabic cuisine opening across Kerala. Rise of group dining culture is another reason for Mandi popularity. Mandi is served in large platters, making it ideal for communal dining, which resonates with Kerala's social culture. Unlike traditional biryani, which is heavily spiced, Mandi's milder seasoning (saffron, dry fruits, and slow-cooked meat) appeals to those looking for a different flavour experience.

Kerala exemplifies a dual trend of preserving tradition while embracing global cuisines. Kerala's culinary landscape is increasingly shaped by the parallel forces of Gastro-Nationalism and Culinary Cosmopolitanism, creating a dynamic interplay between tradition and modernity. Gastro-Nationalism in the state finds expression in the resurgence of traditional meals like the Sadhya, an elaborate vegetarian feast typically served during festivals and rituals. Its reemergence in weddings, religious functions, and even modern restaurants is part of a broader movement to reclaim indigenous culinary heritage. Ayurvedic diets, rooted in holistic health and seasonal balance, are being revived not only in domestic kitchens but also in health resorts and wellness retreats. Millets, once a marginalized staple, are making a robust comeback as part of diabetes-friendly and eco-conscious diets. These shifts underscore a growing pride in native food wisdom and a deliberate distancing from heavily processed or imported ingredients.

At the heart of this Gastro-Nationalist revival is the celebration of local ingredients that were once sidelined in the face of imported tastes. Jackfruit, now branded as a superfood, is used in everything from chips to curries, while coconut continues to hold its position as a culinary cornerstone. Traditional rice varieties like Njavara, Chomala, and Kuruva are being reintroduced into agricultural and dietary practices for their nutritional and medicinal value. These efforts are supported by state-sponsored initiatives and grassroots movements that aim to document and preserve culinary knowledge passed down through generations. Food heritage mapping projects, recipe archiving programs, and cultural food festivals further institutionalize this return to roots, turning food into a vehicle of cultural pride and resistance to homogenized globalization.

Conversely, Culinary Cosmopolitanism thrives in Kerala's urban and diasporic spaces, fueled by transnational migration, media exposure, and evolving consumer preferences. The influx of Gulf-returned migrants has brought with it a love for Arabic-inspired dishes like Mandi Biryani, Shawarma, and Kuboos-based meals. These items are not just available in specialty restaurants but have also made their way into roadside eateries and home kitchens, often modified to suit the Malayali palate. Indo-Chinese food, with its bold use of soy, chili, and vinegar, enjoys enduring popularity, symbolizing a hybrid culinary identity that merges local and foreign techniques. Western fast-food chains, too, have found a stronghold, particularly among the youth, who are influenced by global pop culture and social media trends. Additionally, North Indian dishes like Chole Bhature and Paneer Butter Masala and Thai staples like Green Curry and Tom Yum Soup are being customized with Kerala spices, creating new gastronomic expressions. This culinary openness does not dilute Kerala's identity but rather adds complexity to its food culture, illustrating how global and local influences can coexist and enrich one another.

The post-liberalization period introduced a dramatic shift in food patterns. Fast foods, sugar-laden snacks, and hybrid global dishes displaced local staples. As Popkin's theory predicts, this transition correlates with a rise in lifestyle ailments. Yet, the health crisis has catalyzed a cultural awakening: a growing section of Kerala's population is returning to its culinary roots—not as nostalgia but necessity. This resurgence is evident in diverse spaces. Ayurvedic cafés offering tailored thalis, online content promoting seasonal and sattvic recipes, and government programs that endorse local grains and traditional breakfasts. Ayurveda's dietary system, based on dosha balancing and digestive fire (Agni), emphasizes compatibility, seasonality, and mindful eating. Its resurgence reflects a desire for food that heals rather than harms.

In this context, Gastro-Nationalism takes on new meaning. By promoting Ayurvedic food as a marker of Kerala's cultural and medicinal heritage, food becomes a medium of selfdefinition and resistance. It's a way of saying: "Our roots are relevant." Unlike the passive consumption of globalized food, eating traditionally becomes an act of cultural participation and political assertion. Simultaneously, Culinary Cosmopolitanism has not been discarded but reinterpreted. Dishes like millet pizza or jackfruit tacos marry ancestral ingredients with contemporary formats, showing that Kerala's foodscape can be both rooted and adaptive. These innovations are not betrayals of tradition but testaments to its vitality.

Furthermore, the revival is being institutionalized. Ayurveda-based meal plans are entering urban households; state nutrition missions advocate millets and native greens; and health influencers narrate personal transformations through dietary return. These movements are not isolated trends but signal a collective, conscious pivot toward sustainable, culturally aligned food systems.

With growing health awareness, many people are now looking to return to traditional and organic foods. Ayurveda, the ancient Indian system of medicine, emphasizes a holistic approach to health, integrating the body, mind, and spirit. It promotes preventive healthcare, natural healing, and lifestyle balance through diet, herbal remedies, yoga, and mental wellbeing practices. Some emerging trends include: Farm-to-Table & Organic Eating: A renewed interest in unpolished rice, jackfruit-based dishes, and millet-based meals. And Promotion of Ayurveda-Based Diets, herbal drinks, traditional snacks, and immunity-boosting recipes are regaining popularity.

Ayurveda and Its Role in Holistic Well-Being

Ayurveda, the ancient Indian system of medicine, emphasizes a holistic approach to health, integrating the body, mind, and spirit. It promotes preventive healthcare, natural healing, and lifestyle balance through diet, herbal remedies, yoga, and mental well-being practices. Ayurveda provides a comprehensive system for holistic well-being, integrating diet, lifestyle, natural therapies, and mental health practices. By following Ayurvedic principles, individuals can achieve long-term health, emotional balance, and a harmonious connection with nature.

Ayurveda promotes a holistic approach to eating, centering around a Sattvic diet that prioritizes balance, purity, and digestive health. This system encourages the consumption of fresh, seasonal, and natural foods such as whole grains, fruits, vegetables, nuts, and legumes, all believed to nourish both body and mind. The use of digestive spices like turmeric, cumin, ginger, and coriander not only enhances flavor but also aids detoxification and gut function. Equally important is the practice of mindful eating—consuming food in a calm state, in moderation, and at proper times—to maintain internal harmony and avoid aggravating the body's doshas or energies.

Ayurveda classifies individuals into three primary doshas—Vata, Pitta, and Kapha— each representing different elements and bodily traits. Diets are tailored to each dosha to maintain equilibrium. Vata types, associated with air and space, benefit from warm, grounding foods like soups, dairy, and cooked grains. Pitta, governed by fire and water, requires cooling and hydrating ingredients such as coconut, cucumbers, and ghee to soothe their intense metabolism. Kapha, linked to earth and water, thrives on light, spicy, and warming foods like ginger, leafy greens, and legumes to combat sluggish digestion and maintain vitality.

Ayurveda also emphasizes certain superfoods that contribute to holistic well-being by supporting digestion, immunity, and detoxification. Ghee and ginger are considered powerful digestive agents, while herbs like cumin, coriander, and fennel prevent bloating and acidity.

For immunity and vitality, turmeric is prized for its anti-inflammatory properties, tulsi for respiratory health, and amla for its high vitamin C content. These ingredients are not only traditional but are increasingly recognized in modern nutrition for their wide-ranging benefits.

In terms of detox and energy, Ayurveda recommends hydrating and cleansing foods such as coconut water, lemon-honey drinks, and leafy greens that cleanse the liver and purify the blood. Foods like almonds, dates, figs, and millets provide sustainable energy, healthy fats, and essential nutrients to support overall strength and stamina. Through this framework, Ayurveda presents a timeless, adaptable guide to eating that harmonizes body, mind, and environment—offering both prevention and healing in an age of lifestyle-related ailments.

Conclusion

Kerala stands at a culinary and cultural crossroads where the old and new coexist in dynamic tension. The reemergence of traditional foodways and Ayurveda is not a romanticized return but a strategic move toward health, heritage, and sustainability. Cultural food theories illuminate how this shift is both symbolic and substantive.

The revival of Kerala's traditional food practices is increasingly shaped by the farm-totable movement, which emphasizes the use of organic, locally-sourced ingredients. This shift marks a conscious departure from industrialized food systems and reflects a growing awareness about the environmental and health impacts of processed foods. Small-scale farmers, local markets, and home gardens are gaining renewed importance as people seek fresher, chemicalfree produce. This resurgence aligns closely with traditional agrarian practices in Kerala, where food was once inherently seasonal and regional. By returning to these roots, the community not only ensures nutritional integrity but also supports local economies and fosters sustainable food habits.

Simultaneously, there is a noticeable rise in Ayurveda-inspired diets, which promote sattvic (pure), plant-based meals, herbal decoctions, and dosha-specific eating patterns tailored to individual constitutions. Health-conscious consumers are embracing this holistic model, incorporating Ayurvedic superfoods like turmeric, ghee, tulsi, and amla into their daily routines for their immunity-boosting and therapeutic properties. Restaurants and wellness cafes now offer curated Ayurvedic menus, while wellness influencers and nutritionists advocate for mindful, Ayurvedic eating. These trends reflect a broader cultural and health-oriented shift—a future in which food is not only a source of nourishment but also a personalized tool for wellness, rooted in age-old indigenous wisdom.

In the face of rising health crises and cultural dilution, going back to the roots offers Kerala a path forward. Food becomes not just nourishment but narrative, aligning personal wellness with collective memory. As the state reclaims its edible past, it crafts a resilient future—where eating locally and living seasonally are not just Ayurvedic prescriptions but ethical, ecological, and existential choices.

Bibliography
- Appadurai, A.. How to Make a National Cuisine: Cookbooks in Contemporary India.
- Comparative Studies in Society and History, 30(1), 1988. 3–24.
- Bourdieu, P. *Distinction: A Social Critique of the Judgement of Taste.* Harvard University Press. 1984.

- Douglas, M. Deciphering a Meal. Daedalus, 101(1), 1972. 61–81.
- Fischler, C.. Food, Self and Identity. Social Science Information, 27(2), 1988. 275–292.
- Lévi-Strauss, C. *The Raw and the Cooked.* Harper & Row. 1969.
- Popkin, B. M.. "Global Nutrition Dynamics: The World is Shifting Rapidly toward a Diet Linked with Noncommunicable Diseases". *American Journal of Clinical Nutrition*, 84(2), 2006. 289–298.
- Sreekumar, T. T. "Ayurveda and Health Tourism in Kerala: Regional Traditions and
- Cultural Politics". J. R. Taylor. Ed. *Health and Healing in South Asia.* Routledge, 2020
- Menon, M.." Rediscovering Millets: The Return to Ancient Grains in Modern Kerala" *Indian Journal of Traditional Knowledge*, 18(3), 2019. 472–480.
- Government of Kerala. *Kerala State Health Policy. Department of Health and Family Welfare.* 2022.
- Department of AYUSH. *Guidelines on Ayurveda-Based Diet and Nutrition.* Ministry of AYUSH, Government of India. 2021.
- Kumar, R. "Gastro-Nationalism and Culinary Revival in South India". *Food and Culture*, 12(4), 2021. 102–117.
- Rajagopal, A. . "Sattvic Diets in Contemporary Ayurveda: A Cultural and Nutritional
- Perspective". *International Journal of Ayurvedic Medicine*, 14(2), 2023. 88–96.

Ethnobotany: Study of Indigenous Knowledge, with Special Reference to Tribal Medicine

*¹Dr. Sushama Raj R V

*¹Associate Professor of Botany, VTM NSS College,
Dhanuvachapuram,
Affiliated under University of Kerala (Corresponding author)
drsushamarajrv@gmail.com

²Rajeswari L

²Assistant Professor of Botany (on contract),
VTM NSS College, Dhanuvachapuram,
Affiliated under University of Kerala

Abstract

Ethnobotany is the examination of how individuals from a specific culture and location utilize indigenous plants. Ethnobotany is an interdisciplinary domain that examines the interactions between humans and plants, situated at the convergence of natural and social sciences. It emphasizes the conventional understanding of the utilization, management, and perception of plants within human society. Ethnobotany amalgamates insights from botany, anthropology, ecology, and chemistry to examine plant-related traditions across cultures. Researchers in this domain chronicle and study the utilization of native flora by diverse communities for purposes such as medicine, food, religious practices, intoxicants, construction materials, fuels, and textiles. Richard Evans Schultes, commonly known as the "father of ethnobotany," offered a preliminary definition of the field.

All ancient civilizations developed their own medicinal systems; however, India is particularly noted for its traditional medicinal systems—Ayurveda, Siddha, and Unani—which are regarded as the most systematic and holistic approaches. The use of traditional medicine in India has persisted from ancient times to the present, although documentation has been inadequate. Numerous communities in India engage in the practice of traditional medicine through various methods. The subcontinent serves as a significant repository of medicinal plants utilized in traditional medical treatments. In India, approximately 70% of the rural population relies on traditional medicine, while in Western countries, around 40% of individuals utilize herbal medicines exclusively. Numerous remote tribes depend entirely on the traditional healing system previously mentioned, which is further complemented by Tibetan traditional medicines and localized medicinal knowledge specific to each tribe. However, the potential of tribal traditional medicine has not yet been fully explored (Soumitra and Saikat, 2023).

History

The word ethnobotany was introduced by the early 20th-century botanist John William Harshberger. The origins of ethnobotanical science can be found in ancient Sanskrit, Arabic literature, Greek texts, ethnographic studies, and travelogues. Extensive ethnobotanical knowledge has been present in India since ancient times. Numerous applications of plants are referenced in ancient Indian Sanskrit literature, such as the Rigveda, Atharvaveda, Upanishads, Mahabharata, and Puranas. These encompass plants utilized in religious rituals, medicinal applications, agricultural implements, sustenance, and fuel. A compilation of significant Indian treatises is provided across two Vedic periods. Rigveda and Atharvaveda Charaka Samhita has 148 medicinal plants, while 400-450 medicinal plants are documented overall. Pent-s'ao, the herbal compendium authored by Emperor Shah Nung, has references to 365 medicinal substances. Reports indicate that the Egyptians routinely utilized hundreds of medications, including significant species such as henbane, pomegranate, opium, poppy, aloe, and onion. Ethanobotany has recently emerged as a significant scientific discipline. The primary concerns of ethnobotanical studies pertain to the relationship between plants and humans, particularly the stewardship of plant variety by indigenous societies and the conventional utilization of medicinal plants.

TYPES OF ETHNOBOTANICAL STUDY

1. **Medicinal Ethnobotany:-** The medical use of ethnobotany reveals a wealth of traditional healing traditions. As various cultures refined their expertise over years, numerous herbal medicines developed. For ages, established methods of medical ethnobotany have significantly influenced modern medicine, guiding the search for remedies for many physical and mental disorders. These ancient traditions have produced life-saving drugs

and innovative cures, significantly influencing global healthcare.

2. **Ethnopharmacology:-**Ethnopharmacology investigates the scientific study of traditional medical practices among indigenous cultures. Through the analysis of the chemical makeup of plants and their possible therapeutic benefits, researchers reveal intriguing avenues for novel pharmacological discoveries. This branch establishes a connection between traditional healing methods and contemporary medical research, providing optimism for innovative therapies and possible cures. Each discovery in ethnopharmacology facilitates significant breakthroughs in healthcare and underscores the necessity of preserving the rich legacy of herbal treatments.

3. **Ritual and Ceremonial Ethnobotany:-**The realm of ritualistic and ceremonial use of ethnobotany invites exploration of sacred plants and their spiritual importance. Various cultures worldwide integrate particular plants into their religious rituals, establishing a deep bond between humans and the divine. This branch reveals the significant symbolism associated with these sacred plants and their essential significance in cultural traditions, providing insight into the spiritual essence that enhances human existence. The use of sage in Native American ceremonies and the importance of sacred trees in ancient rites exemplify the deep connection between flora and spirituality, highlighting the profound enigma of existence.

4. **Culinary Ethnobotany:-**Explore the compendium of gastronomic ethnobotanical plants, revealing the depth of cultural gastronomy. This intriguing field examines the varied cuisines globally, fundamentally based on the utilization of indigenous flora. Culinary ethnobotany, in addition to stimulating taste senses, uncovers the nutritional

advantages of traditional foods, providing significant insights into health, wellness, and dietary patterns that have supported generations. Culinary ethnobotany highlights the ethnic diversity that enriches our diets, from India's exquisite spices to Latin America's robust grains.

5. **Historical Ethnobotany:-**In the intriguing domain of historical medical ethnobotany, the botanical knowledge of past civilizations is vividly expressed. Researchers examine historical records to uncover the lifestyles, foods, and medicinal practices of our ancestors, mapping the progression of human-plant interactions over time. This branch venerates the sagacity of our ancestors and derives motivation from their sustainable methodologies to tackle modern environmental issues. The remnants of ancient customs, recorded in historical ethnobotanical plant lists, underscore the enduring relationship between humanity and the natural environment—an association that persists as a stabilizing force in an increasingly dynamic contemporary context.

Ethnomedicine

Ethnomedicine is the examination or comparison of traditional medicine derived from bioactive substances in flora and fauna, utilized by diverse ethnic groups, particularly those with limited access to Western medicine, such as indigenous populations. Ethnomedicine is occasionally employed as a synonym for traditional medicine. Ethnomedical research is interdisciplinary, employing the methodologies of ethnobotany and medical anthropology in the examination of indigenous remedies. The medicinal traditions it examines are frequently maintained solely through oral transmission. Alongside flora, several traditions involve notable relationships with insects in the Indian Subcontinent, Africa, and other regions worldwide (Acharya *et al.*, 2008).

Ethnomedicine is a sub-discipline of medical anthropology focused on the examination of traditional remedies, encompassing both those with documented sources (such as Traditional Chinese Medicine and Ayurveda) and those whose knowledge and practices have been orally transferred through generations. Although ethnomedical studies primarily concentrate on indigenous perceptions and applications of traditional medicines, another impetus for this research is the discovery and development of pharmaceuticals. Prominent pharmaceuticals like digoxin, morphine, and atropine are derived from foxglove, opium, and belladonna, respectively.

Ethnomedical research in this century has resulted in the creation of significant pharmaceuticals, including reserpine (a hypertension treatment), podophyllotoxin (the foundation of a crucial anti-cancer medication), and vinblastine (utilized in the treatment of specific cancers). In the scientific domain, ethnomedical research are typically defined by a robust anthropological perspective or a pronounced biomedical focus, especially in drug discovery initiatives. Anthropological studies concentrate on the perception and contextual application of traditional remedies, whereas biological efforts typically emphasize the identification of medicinal compounds, such as the anti-HIV/AIDS compound prostratin.

Indigenous Communities and Tribal Medicine

India possesses a substantial repository of tribal knowledge. Indigenous and tribal populations reside in proximity to forest flora and fauna for sustenance, medicinal resources, and materials for constructing shelters. The Himalayan and Aravalli ranges, Eastern Ghats, and the biodiversity hotspots in Arunachal Pradesh and the Western Ghats exhibit extensive forest cover. Traditionally, tribal knowledge has been transmitted across generations via oral communication. The tradition is declining as the new generation distances itself from tribal cultures and habitats. The nomenclature and terminology employed by tribes for flora are specific to their regions, and the unrecorded knowledge will be irretrievably lost. It has been observed that tribal communities often exhibit reluctance in disclosing their knowledge. Anthropologists must establish rapport with individuals to obtain information. Research indicates that the plants utilized by the tribal population differ from AYUSH medicinal plants in their application for treating specific diseases. Rauwolfia serpentina serves as an antidote for snakebite within tribal communities, while reserpine, derived from the plant, is utilized in the treatment of hypertension. In tribal tradition, numerous fresh leaves are processed into a paste or crushed to obtain juice for the treatment of various ailments.

The tribal population of India is distributed along a belt that extends from eastern Gujarat and Rajasthan in the west to the eastern states of Nagaland and Mizoram. This area is referred to as the 'tribal belt.' The 'tribal belt' corresponds approximately to three regions. The western region encompasses eastern Gujarat, south-eastern Rajasthan, north-western Maharashtra, and western Madhya Pradesh, and is predominantly inhabited by Indo-Aryan speaking tribes such as the Bhils. The central region encompasses eastern Maharashtra, Madhya Pradesh, western and southern Chhattisgarh, northern and eastern Telangana, northern Andhra Pradesh, and western Odisha, and is predominantly inhabited by Dravidian tribes such as the Gonds and Khonds. The eastern belt, centered on the Chhota Nagpur

Plateau in Jharkhand and surrounding regions of Chhattisgarh, Odisha, and West Bengal, is primarily inhabited by Munda tribes, including the Hos and Santals (Soumitra and Saikat, 2023).

Kerala Scenario

Kerala is recognized for its abundant greenery and possesses a significant tribal heritage. Referred to as Adivasi, these indigenous communities have inhabited the forests and mountains of the Western Ghats for centuries. The ranges adjacent to Karnataka and Tamil Nadu support an ecosystem for tribal communities that have historically engaged in sustainable living practices in alignment with nature. The Indian government has designated these indigenous communities as "Scheduled Tribes," providing them with specific rights and benefits aimed at addressing historical inequalities. The Kerala Institute for Research Training and Development Studies of Scheduled Castes and Scheduled Tribes (KIRTADS) recognizes 36 unique Scheduled Tribes in the state. The Scheduled Tribes Development Department further classifies these groups into three subcategories: Particularly Vulnerable, Marginalised, and Minorities. The 2011 census indicates that the tribal population of Kerala comprises approximately 1.5 percent of the total population, amounting to around 484,839 individuals. Wayanad district has the highest concentration of tribal communities, succeeded by Idukki, Palakkad, Kasargod, and Kannur. The major tribal communities in Kerala include Paniyar, Irular, Kattunaikan, Oorali, and Adiyar.

Tribes possess extensive knowledge of traditional medicine for a variety of diseases. The tribes' superstitious beliefs prevent them from disclosing their medicinal secrets to others. Consequently, a specific medical treatment is lost upon the death of the knowledge holder. Several independent studies conducted among different tribal communities of India helped the scientists to identify different medicinal plants and their uses administrated by these indigenous communities.

Conservation of Indigenous Knowledge

Preserving traditional knowledge serves to protect historical insights while simultaneously enhancing future prospects. This body of knowledge includes medicinal practices utilizing local plants, sustainable agricultural techniques, environmental conservation methods, as well as intricate crafts, folklore, and rituals. Traditional knowledge is fundamentally intertwined with the environment, history, and spirituality of its respective community, providing distinctive perspectives and enduring solutions. Traditional medicinal knowledge serves as the basis for numerous contemporary pharmaceuticals. Indigenous communities have historically utilized local flora and natural resources for the treatment of illnesses and the maintenance of health. Fabricant and Farnsworth (2001) emphasize that numerous modern pharmaceuticals originate from traditional remedies, highlighting the necessity of preserving this knowledge for current and future medical advancements. The erosion of traditional medicinal knowledge may result in the loss of potential disease cures and weaken healthcare systems, particularly in rural and indigenous populations.

Traditional knowledge possesses considerable economic potential, especially in sectors such as pharmaceuticals, agriculture, and tourism. Safeguarding this knowledge guarantees that indigenous communities reap the benefits of their intellectual property, rather than facing exploitation by external entities. According to Graham and Shibata (2007), legal frameworks and intellectual property rights are essential for preventing biopiracy and guaranteeing equitable compensation for the utilization of traditional knowledge. This economic empowerment can facilitate community development and the revitalization of traditional practices. Protecting traditional knowledge encompasses both ethical and legal dimensions. Recognizing and respecting the rights of indigenous and local communities to their knowledge and practices is essential. International agreements, including the Convention on Biological Diversity (CBD), establish frameworks for the protection and equitable distribution of benefits arising from traditional knowledge (CBD, 1992).

To preserve, safeguard, and enhance awareness of traditional knowledge, India has implemented several comprehensive measures, including the establishment of the Traditional Knowledge Digital Library (TKDL), the Ministry of AYUSH, the Biological Diversity Act (2002), the National Intellectual Property Rights Policy (2016), and the Indian Knowledge Systems (IKS) innovative cell under the Ministry of Education. The incorporation of ancient knowledge into educational curricula, the legal structures governing intellectual property rights, and the encouragement of multidisciplinary research further bolster these preservation initiatives. India integrates modern technology with traditional methods, so safeguarding its cultural heritage while promoting innovation and enhancing global acknowledgment of its indigenous expertise. These endeavours are essential for preserving the extensive repository of traditional knowledge for the benefit of future generations, both in India and worldwide (Sudipta Shee, 2022).

References

- A.G. Devi Prasad., T. B. Shyma, and M. P. Raghavendra. (2013). Plants used by the tribes for the treatment of digestive system disorders in Wayanad district, Kerala. Journal of Applied Pharmaceutical Science. 3(08): 171-175.
- Acharya, Deepak and Shrivastava Anshu. (2008). *Indigenous Herbal Medicines: Tribal Formulations and Traditional Herbal Practices*. Aavishkar Publishers Distributor, Jaipur / India. ISBN 978-81-7910-252-7, 440.
- Bhosle S. V., Ghule V. P., Aundhe D. J. and Jagtap S. D. (200)9. Ethnomedical Knowledge of Plants used by the Tribal people of Purandhar in Maharashtra, India. Ethnobotanical Leaflets. 13. 1353-1361.
- Fabricant, D. S., & Farnsworth, N. R. (2001). The value of plants used in traditional medicine for drug discovery. Environmental Health Perspectives, 109(Suppl 1), 69-75.
- Graham, J., & Shibata, A. (2007). Using intellectual property rights to preserve and promote traditional knowledge: The benefit-sharing model. The Innovation Journal: The Public Sector Innovation Journal, 12(3), 1-16.
- https://forest.kerala.gov.in/en/indigenous-communities-of-kerala/#:~:text=Ulladan:%20The%20Ulladan%2C%20also%20known%20as%20Nayadi,is%20a%20significant%20challenge%20for%20the%20community.
- https://www.euroschoolindia.com/blogs/ethnobotany-meaning-types-and-applications/
- Husen, A. (Ed.). (2022). Herbs, Shrubs, and Trees of Potential Medicinal Benefits (1st ed.). CRC Press. https://doi.org/10.1201/9781003205067

- Inayat Ur Rahman and Aftab Afzal and Zafar Iqbal and Farhana Ijaz and Niaz Ali and Muzammil Shah and Sana Ullah and Rainer W. Bussmann. (2019). Historical perspectives of ethnobotany. Clinics in Dermatology. 37(4). 382-388. https://doi.org/10.1016/j.clindermatol.2018.03.018
- Mondal, Soumitra & Bhattacharya, Saikat. (2023). Tribal Medicine of India: an evolving ancient tradition. 1. 3-5.
- Pramukh, K.E. &Palkumar, P.D.S. (2006). Indigenous Knowledge: Implications in Tribal Health and Disease. Studies of Tribes and Tribals. 4. 1-6. https://doi.org/10.1080/0972639X.2006.11886531
- Rai, M., Bhattarai, S., & Feitosa, C.M. (Eds.). (2021). Ethnopharmacology of Wild Plants (1st ed.). CRC Press. https://doi.org/10.1201/9781003052814
- Ram Prakash, (2015). Medicinal Plants Used By Tribal Communities: A Study of Uttarakhand Himalayan Region. International Journal of Humanities and Social Science Invention. 4(21). 55-61.
- Sharad Kumar Singhariya., (2023). An Introduction to Ethanobotany, Concept, History Importance and Scope. International Journal of Novel Research and Development. 8(10), 2456-4184.
- Sharangi, A.B., & Peter, K.V. (Eds.). (2022). Medicinal Plants: Bioprospecting and Pharmacognosy (1st ed.). Apple Academic Press. https://doi.org/10.1201/9781003277408
- Shyma T.B. and Devi Prasad A.G. (2012). Traditional use of Medicinal Plants and its status among the tribes in Mananthavady of Wayanad District, Kerala. World Research Journal of Medicinal &

Aromatic Plants. 1(2): 22-26.

- Sivaraman M A. (2014). Trible Medicine in Western Ghats. Shanlax International Journal of Arts, Science & Humanities. 1(4). 39-41.
- Sudipta Shee, (2022). Traditional Knowledge Preservation: An Overview of Strategies and Challenges. Indian Knowledge System. 167-171. ISBN: 978-81-974990-0-5
- Uniyal SK, Singh KN, Jamwal P, Lal B. (2006). Traditional use of medicinal plants among the tribal communities of Chhota Bhangal, Western Himalaya. J EthnobiolEthnomed. 2 (14). doi: https://doi.org/10.1186/1746-4269-2-14 .

The Feminine Legacy: Women's as Cultural Archivists in Sustaining India's Traditional Knowledge

Dr Seema Rajan S

Assistant Professor, Department of English
NSS College, Nilamel, Kollam

Abstract

Women have been pivotal in preserving and transmitting India's traditional knowledge systems, serving as custodians of cultural, ecological, and medicinal wisdom. Despite their indispensable contributions, their roles have often been marginalized in mainstream narratives. This paper examines the multifaceted involvement of women in sustaining traditional knowledge, particularly in agriculture, healthcare, crafts, and oral traditions. Through an analysis of historical evidence and case studies, such as the seed-saving practices of rural women and the artistic contributions of the Devadasi tradition, the study underscores the critical role of women in safeguarding India's intellectual and cultural heritage. It also highlights the challenges they face, including lack of recognition, documentation gaps, and socio-economic marginalization. By recognizing and integrating women's knowledge, India can create a more inclusive and sustainable framework for preserving its rich heritage.

Keywords: Indian knowledge system, feminism, globalisation, urbanisation, marginalisation, cultural archivists, empowerment.

Introduction

Traditional knowledge systems in India encompass a diverse range of practices, including agriculture, medicine, arts, and crafts, which have been preserved and transmitted across generations. Women have played a central role in sustaining these systems, acting as primary caregivers, educators, and practitioners within their communities. Through oral traditions, rituals, and daily practices, women have safeguarded indigenous wisdom, such as herbal medicine, sustainable farming techniques, and traditional crafts. However, their contributions are often overlooked, and they face significant challenges, including lack of recognition, documentation gaps, and socio-economic marginalization, exacerbated by rapid modernization. This paper explores the pivotal role of women in preserving traditional knowledge systems, highlighting their contributions and the obstacles they encounter. By examining case studies and historical evidence, the study underscores the need for empowering women as key stakeholders in the revitalization of India's intellectual and cultural heritage, ensuring its continuity and relevance in a globalized world.

Historical Context

Women have been central to preserving and transmitting India's traditional knowledge systems, encompassing agriculture, medicine, arts, and crafts. As primary caregivers, educators, and practitioners, they have safeguarded cultural, ecological, and practical wisdom through oral narratives, rituals, and daily practices. In ancient and medieval India, women played pivotal roles in sustaining these systems, from ecological practices like sustainable farming to artistic traditions such as weaving and folk art. As Shakti M. Gupta states: "Women played a central role in... the preservation of knowledge. It required a deep understanding of religious texts, herbal medicine, and culinary arts, making women key figures in the transmission of cultural and spiritual knowledge" (45). However, colonialism disrupted these roles, marginalizing women's contributions and devaluing indigenous knowledge. Despite this, women demonstrated resilience, adapting to socio-political changes while continuing to preserve traditional practices.

Post-independence, women's contributions have persisted, particularly in rural and tribal communities, where they remain key custodians of cultural and ecological wisdom. Yet, their roles are often overlooked in mainstream narratives, hindering their empowerment and the revitalization of traditional knowledge systems. To address this, there is a need for formal recognition, documentation, and integration of women's knowledge into policy-making and educational frameworks. Empowering women as cultural archivists not only honours their contributions but also ensures the preservation and revitalization of India's intellectual and cultural heritage for future generations.

Women as Knowledge Keepers

In ancient and medieval India, women played a central role in preserving and transmitting cultural, ecological, and practical knowledge. During the Vedic Period, women such as Gargi and Maitreyi emerged as prominent scholars and philosophers, actively engaging in intellectual discourses and contributing to Vedic literature. Their participation in philosophical debates and scholarly contributions underscores the significant role women played in India's early intellectual traditions, challenging the misconception that women were excluded from such spheres. Swami Prabhavananda confirms:

In ancient India, women were not only permitted but encouraged to pursue knowledge and spirituality. The Rigveda mentions women scholars like Gargi and Maitreyi, who engaged in philosophical debates and contributed to the intellectual tradition of the time. These women were highly revered and respected for their wisdom and played a significant role in shaping the Indian knowledge system. (112)

The elite scholarly class, women in households and communities served as primary educators, responsible for imparting traditional practices such as cooking, farming, herbal medicine, and crafts to younger generations. For instance, women in rural households have historically preserved heirloom seed varieties and sustainable farming techniques, ensuring the continuity of these practices across generations. Their role as transmitters of practical knowledge was indispensable to the survival, prosperity, and cultural continuity of their communities. This historical evidence highlights the resilience and intellectual contributions of women, emphasizing their integral role in India's traditional knowledge systems.

Women also made significant artistic and spiritual contributions, preserving and transmitting performing arts such as dance, music, and storytelling, often linked to religious and cultural rituals. The Devadasi tradition, for instance, involved women in the preservation of classical dance and music, embedding these art forms within the spiritual and cultural fabric of society. Through their artistic endeavours, women not only sustained cultural traditions but also enriched India's artistic heritage, ensuring its survival for future generations. Vijaya Nagarajan sums up the role of women in India by stating, "Women have been the keepers of India's oral traditions, passing down folk tales and epics that embody the wisdom of generations". (33)

Colonial Impact

The advent of colonial rule in India significantly disrupted traditional knowledge systems, disproportionately affecting women, who were primary custodians of these traditions. Colonial policies prioritized Western education and scientific knowledge, systematically devaluing indigenous practices. This marginalization eroded women's authority in domains like traditional healthcare, agriculture, and crafts, often dismissed as unscientific or backward. The colonial focus on male-dominated institutions further side-lined women's contributions, relegating their knowledge to societal margins.

The introduction of industrial agriculture, modern medicine, and mass-produced goods accelerated the decline of practices like organic farming, herbal medicine, and handicrafts, predominantly managed by women. For instance, traditional midwives were displaced by modern healthcare, and handwoven textiles were replaced by factory goods, disrupting women's economic and cultural roles. This erosion marginalized women and led to the loss of invaluable cultural and ecological wisdom preserved by them for centuries.

The historical context of women's roles in preserving and transmitting traditional knowledge in India highlights their resilience and adaptability amid significant challenges.

However, colonial disruptions and the marginalization of indigenous knowledge systems disproportionately impacted women, eroding their authority and devaluing their contributions. Recognizing and revitalizing women's roles in traditional knowledge systems is crucial for preserving India's heritage and fostering an inclusive future. Addressing historical and systemic barriers can empower women as key stakeholders in safeguarding traditional knowledge, ensuring its continuity and relevance. Mira Roy comments:

In traditional Indian society, women were the custodians of household knowledge, including culinary arts, childcare, and home remedies. This practical wisdom, passed down through generations, formed an essential part of the Indian knowledge system, ensuring the well-being of families and communities. (79)

Post-Independence Era

Despite India's independence in 1947 and subsequent modernization, women in rural and tribal communities remain vital in preserving and transmitting traditional knowledge. They are primary custodians of practices like seed preservation, sustainable farming, and herbal medicine, safeguarding indigenous seed varieties and agricultural techniques that ensure biodiversity and food security. Their expertise in local ecosystems and sustainable practices has been crucial for ecological balance and resilience. Vandana Shiva aptly states:

Women in India have preserved traditional ecological knowledge, such as sustainable farming practices and water conservation techniques. Their understanding of the natural world, passed down through generations, has been crucial in maintaining the balance between humans and nature. Women often lead community efforts to protect forests, rivers, and farmland, drawing on this traditional wisdom. Their role as cultural archivists in this field has been vital in promoting environmental sustainability. (137)

Women have also demonstrated cultural resilience by sustaining folk arts, crafts, and oral traditions amid globalization. For example, Mishing women in Assam preserve traditional weaving techniques, producing handloom textiles that reflect their cultural heritage, while women storytellers and folk artists keep oral histories and myths alive, ensuring the continuity of cultural narratives in a rapidly modernizing world.

However, modernization poses significant threats to the preservation of traditional knowledge systems, marginalizing women who uphold them. While modernization has brought progress, it has also eroded indigenous practices, leaving women in rural and tribal communities undervalued and unrecognized. Economic marginalization exacerbates these challenges, as women struggle to earn fair market value for their crafts, agricultural produce, or traditional healthcare services. Limited access to education and resources further restricts their ability to adapt, while modern education systems, prioritizing Western knowledge, often neglect traditional practices. This leaves women without platforms to sustain and transmit their wisdom.

In short, the post-independence era has been marked by both resilience and struggle for women in rural and tribal communities. While they continue to preserve traditional knowledge, modernization threatens its survival. Addressing these challenges requires recognizing women's contributions, providing economic and educational support, and integrating traditional knowledge into contemporary frameworks. Empowering women as key stakeholders can preserve India's cultural heritage and ensure a more inclusive, sustainable future. Pupul Jayakar affirms:

Women in rural India have been the custodians of folk-art forms like Madhubani and Warli painting, ensuring the survival of these traditions through their creativity and dedication. These art forms, often created by women, depict scenes from daily life, mythology, and nature, serving as a visual record of India's cultural heritage and tradition. Women pass down these artistic skills to their daughters, ensuring that the traditions remain alive. Their work as cultural archivists has been vital in preserving India's rich artistic legacy. (82)

Challenges Faced by Women

Women in India, despite their pivotal role in preserving and transmitting traditional knowledge, face significant challenges, including a lack of recognition and documentation gaps. Their contributions to agriculture, healthcare, crafts, and oral traditions are often undervalued and excluded from mainstream narratives and academic discourse. For instance, women's expertise in preserving heirloom seeds and herbal remedies remains largely unacknowledged in formal policies or institutional frameworks. This marginalization not only undermines their role as knowledge custodians but also threatens the preservation of India's intellectual and cultural heritage. Addressing these challenges is essential to empower women and ensure the continuity of traditional knowledge systems. A. K. Ramanujan asserts:

Women in India have been the custodians of folklore, which is an integral part of the Indian Knowledge System. Through songs, stories, and rituals, women have preserved the collective memory of their communities, passing on knowledge of history, medicine, dance, ecology, and ethics. The temple was often the site where these traditions were performed, with women playing a central role in festivals and rituals. These practices were not just cultural expressions but also means of education, ensuring that traditional knowledge was transmitted to future generations. (34)

Women face multifaceted challenges in preserving and transmitting traditional knowledge, including lack of recognition, economic marginalization, modernization's disruptive impacts, and entrenched gender inequality. Addressing these issues requires concerted efforts to document and value women's contributions, provide economic and educational support, integrate traditional knowledge into modern curricula, and challenge societal norms perpetuating gender disparities. Empowering women as key stakeholders in preserving traditional knowledge is essential for sustaining India's cultural heritage and fostering an inclusive, equitable future.

A Silver-lining and Opportunities

Women in rural Rajasthan have played a pivotal role in preserving traditional seed varieties, contributing significantly to biodiversity and food security. Their expertise in heirloom seeds and sustainable farming practices has been instrumental in maintaining agricultural diversity, offering sustainable alternatives to industrial farming methods. Similarly, in Kerala, women practitioners of Ayurveda and traditional medicine continue to provide healthcare using ancient knowledge, particularly in herbal remedies and midwifery. Their contributions have been vital in sustaining traditional healthcare systems, which emphasize holistic well-being and preventive care. These examples highlight the critical role of women in preserving ecological and medicinal knowledge, ensuring the continuity of practices that are both culturally and environmentally sustainable. David Frawley asserts: "The principles of Ayurveda and Yoga, rooted in the Indian Knowledge System, are increasingly being recognized for their scientific validity and relevance in modern healthcare and wellness practices". (78)

In Assam, the Mishing women have sustained traditional weaving techniques, producing exquisite handloom textiles that reflect their cultural heritage while providing economic sustenance for their communities. Their craftsmanship not only preserves a rich artistic tradition but also empowers them economically. Meanwhile, in Tamil Nadu, women storytellers have kept alive folk tales and devotional songs, enriching the region's cultural heritage. Their role as custodians of oral traditions ensures the continuity of cultural narratives that might otherwise be lost in the face of rapid modernization. These examples underscore the multifaceted contributions of women in preserving India's intellectual and cultural heritage, emphasizing the need to recognize and support their roles in sustaining traditional knowledge systems. By empowering women as key stakeholders, India can ensure the preservation and revitalization of its diverse knowledge traditions. Kamaladevi Chattopadhyay opines:

Women in India have been the custodians of craft traditions, which are an integral part of the Indian Knowledge System. From weaving and pottery to embroidery, ayurveda, dance, culinary skills and jewellery-making, women have preserved these skills through generations. The integration of these craft traditions into modern education can help revive India's cultural heritage and empower women economically. (56)

Conclusion

The historical and contemporary roles of women in preserving and transmitting traditional knowledge in India underscore their indispensable contributions to cultural, ecological, and intellectual heritage. From ancient scholars like Gargi and Maitreyi to rural and tribal women safeguarding seeds, herbal medicine, crafts, and oral traditions, women have been the silent yet resilient custodians of India's traditional wisdom. Despite the disruptions caused by colonialism and the challenges of modernization, women have continued to adapt and sustain these knowledge systems, ensuring their survival for future generations. Recognizing and empowering women as key stakeholders in the preservation of traditional knowledge is not only a matter of cultural justice but also a necessity for fostering sustainable development and holistic education. By integrating their knowledge into contemporary frameworks and addressing the systemic barriers they face, India can uphold its rich heritage while paving the way for a more inclusive and equitable future.

Works Cited

- Chattopadhyay, Kamaladevi. *Indian Handicrafts*. Allied Publishers, 1963.
- Frawley, David. *Ayurveda and the Mind: The Healing of Consciousness*. Lotus Press, 1997.
- Gupta, Shakti M. *Plant Myths and Traditions in India*. Brill, 1971.
- Jayakar, Pupul. *The Earth Mother: Legends, Goddesses, and Ritual Arts of India*. Harper & Row, 1990.
- Nagarajan, Vijaya. *Feeding a Thousand Souls*. Oxford University Press, 2019.
- Prabhavananda, Swami. *The Spiritual Heritage of India*.

Vedanta Press, 1979.

- Ramanujan, A.K. *Folktales from India: A Selection of Oral Tales from Twenty-Two Languages.* Pantheon Books, 1991.
- Roy, Mira. *Women in Indian Folklore.* Indian Publications, 1986.
- Shiva, Vandana. *Staying Alive: Women, Ecology, and Development.* Zed Books, 1988.

Temples as Centres of Knowledge: Decolonizing Indian Education through Indian Knowledge Systems

Dr Sooraj Kumar

Associate Professor and Head, Department of English,
St. John's College, Anchal, Kollam, Kerala-691306

Abstract

The colonial legacy in India's education system has systematically marginalized indigenous knowledge systems, creating a disconnect between traditional wisdom and modern academic frameworks. This paper advocates for the decolonization of Indian education by integrating Indian Knowledge Systems (IKS) into mainstream curricula. Through an analysis of historical context, contemporary challenges, and potential strategies, it underscores the significance of IKS in fostering cultural identity, critical thinking, and sustainable development. The study highlights the need to address documentation gaps, academic biases, and curriculum design challenges to ensure effective implementation. It concludes with recommendations for policymakers, educators, and stakeholders to develop an inclusive and holistic education system that revitalizes India's intellectual heritage, ensuring its relevance in a globalized world.

Keywords: Indian knowledge systems, decolonization, cultural identity, vedic knowledge, cultural identity, sacred texts, spiritual pedagogy.

Introduction

India's education system, historically shaped by colonial policies, has predominantly prioritized Western knowledge frameworks, often marginalizing indigenous traditions. The National Education Policy (NEP) 2020 represents a transformative shift by advocating for the integration of Indian Knowledge Systems (IKS) into mainstream education. This paper examines the imperative to decolonize Indian education, exploring the historical marginalization of IKS, the challenges in its integration, and the strategies required to effectively incorporate it into contemporary curricula. As Balachandra Rao states: "The Indian Knowledge System is not just a collection of ancient texts but a living tradition that integrates science, philosophy, art, and ethics into a holistic framework for understanding life and the universe" (45). By addressing these knowledge of mythology, astronomy, and mathematics. Women, particularly those from the Devadasi tradition, played a crucial role in preserving and performing classical arts, ensuring that these traditions were passed down through generations. (23)

Historical Context of Indian Knowledge Systems in Temples

Throughout Indian history, the construction and patronage of temples played a pivotal role in shaping the socio-economic and cultural fabric of society. Funded by kings, merchants, and local communities, temples transcended their religious functions to become centres of learning, governance, and social organization. They served as repositories of diverse knowledge systems, patronizing scholars, artists, and medical practitioners. The integration of religious teachings with secular sciences, such as astronomy, mathematics, and medicine, underscores the multifaceted role of temples in preserving and disseminating knowledge. This synthesis of spiritual and intellectual pursuits highlights the temples' significance as dynamic institutions that contributed to India's cultural and scientific advancements.

One of the most significant aspects of temple culture was their role as centres of learning. Many temples housed *gurukuls* and *pathshalas*, where students studied scriptures, philosophy, linguistics, and sciences under the guidance of scholars. These temple-based institutions were instrumental in preserving oral traditions and transmitting knowledge across generations. Temples also maintained extensive libraries, housing manuscripts on diverse disciplines such as astronomy, medicine, and literature. A notable example is the Saraswathi Mahal Library in Thanjavur, which originated as a temple library and evolved into one of India's most renowned manuscript repositories. Additionally, temple inscriptions served as educational tools, with detailed records of historical events, mathematical calculations, and astronomical data etched onto temple walls. These inscriptions not only documented the past but also functioned as accessible public resources for education and knowledge dissemination. Rabindranath Tagore states:

India's ancient education system was not confined to the four walls of a classroom; it was a way of life. The gurukuls and ashrams were not just schools issues, the study highlights the potential of IKS to foster cultural identity, interdisciplinary learning, and sustainable development within India's educational framework.

Temples as Centres of Knowledge and Learning

Temples in medieval India served not only as places of worship but also as dynamic centres of knowledge and learning. The multifaceted role of temples in preserving and disseminating knowledge across diverse domains, including religion, philosophy, astronomy, mathematics, medicine, and the arts is undeniable. Radhakrishnan S. asserts: "Temples in India were not just places of worship; they were the epicentres of learning, art, and culture. They served as universities where philosophy, astronomy, medicine, and music were taught and preserved" (47).A thorough analysis of historical evidence, architectural features, and inscriptions, highlights how temples functioned as intellectual hubs, fostering cultural and scientific advancements. Their contributions to education, scholarship, and community life, underscores the importance of recognizing and preserving this legacy as an integral part of India's rich intellectual and cultural heritage.

Medieval India saw the flourishing of temple architecture and culture, with temples emerging as epicentres of spiritual, social, and intellectual life. Beyond their religious functions, temples played a pivotal role in fostering education, scholarship, and the arts. Temples as centres of knowledge and learning made significant contributions to the intellectual and cultural landscape of medieval India. By analysing their multifaceted functions, it is acknowledged that temples served as hubs of intellectual activity, preserving and disseminating knowledge across diverse domains such as philosophy, astronomy, mathematics, and the arts. B. N. Goswamy asserts:

Temples in India were not just places of worship but also vibrant cultural hubs where art, music, dance, and literature flourished. They were centres of learning where scholars and artists gathered to study and create. The temple walls, adorned with intricate carvings and inscriptions, served as visual textbooks, conveying but centres of holistic learning where students imbibed knowledge of the self, society, and the universe. This system emphasized the harmony of the individual with nature and the cosmos, a concept that is deeply relevant in today's fragmented world. (56)

The **domains of knowledge preserved and promoted** within temples were vast and interdisciplinary. In the realm of **religion and philosophy**, temples facilitated the study of Vedas, Upanishads, Agamas, and other religious texts. They served as venues for philosophical debates and discourses, fostering intellectual exchange. Temples also played a critical role in **astronomy and mathematics**, with structures like the **Sun Temple at Konark**and the **Chola temples** incorporating advanced astronomical alignments into their architecture. Scholars associated with temples made significant contributions to mathematics, particularly in the development of **trigonometry and algebra**.

Similarly, medicine and Ayurveda flourished within temple complexes, where traditional healing practices were studied and applied. Temples, particularly in Kerala, were renowned for their Ayurvedic treatment centres, serving as hubs for preserving and transmitting medical knowledge across generations. Additionally, temples became epicentres of classical dance, music, and drama, which were integral to religious rituals. They also hosted literary gatherings and devotional poetry compositions, significantly enriching India's artistic and cultural heritage. By functioning as centres of religious, scientific, and artistic learning, temples played a foundational role in preserving and advancing the Indian Knowledge System. Their influence extended far beyond worship, profoundly shaping the intellectual and cultural fabric of society for centuries.

Architectural Features Supporting Learning

The architectural design of Indian temples was instrumental in fostering education and intellectual discourse. Structures such as mandapas and sabha mandapas—large pillared halls—served as venues for teaching, philosophical debates, and cultural performances. These spaces enabled scholars, poets, and artists to engage in knowledge exchange, ensuring that learning remained a core aspect of temple activities. Additionally, temple sculptures and iconography functioned as visual educational tools, with intricate carvings depicting mythological narratives, scientific principles, and moral lessons. Ananda K. Coomaraswamy says: "Temples have been the custodians of India's ancient traditions, preserving the Vedas, Upanishads, and other sacred texts. They were the living bridges between the past and the present" (93). These artistic representations made knowledge accessible to both literate and illiterate audiences, bridging the gap between scholarly and popular understanding. For example, temple reliefs illustrated astronomical concepts, medical practices, and ethical teachings, reinforcing learning through visual and artistic mediums.

Moreover, some temples incorporated astronomical instruments into their designs, facilitating the study of celestial phenomena. The Jantar Mantar in Jaipur, though not a traditional temple, exemplifies how Indian architecture seamlessly integrated scientific inquiry with religious and educational functions. By combining architectural ingenuity with intellectual pursuits, temples served as dynamic centres of learning, where art, science, and spirituality converged. Stella Kramrisch asserts: "Indian temples are not just architectural marvels; they are symbolic representations of the cosmos.They reflect the profound understanding of the universe embedded in the Indian knowledge system" (83). This integration highlights the multifaceted role of temples in advancing the Indian Knowledge System, demonstrating their significance as institutions that transcended mere religious functions to shape the intellectual and cultural fabric of society.

Case Studies of Temple-Based Learning

Several temples in India exemplify the pivotal role of religious institutions in the preservation and dissemination of knowledge. The Nataraja Temple in Chidambaram is renowned for its association with dance and music, serving as a centre for the study of *Natyashastra* and performing arts. Its sculptures and inscriptions preserve the theoretical foundations of classical dance, particularly Bharatanatyam, highlighting the temple's contribution to India's artistic heritage. Similarly, the Brihadeeswarar Temple in Thanjavur, constructed during the Chola dynasty, functioned as a hub of learning where scholars made significant contributions to Tamil literature, astronomy, and architecture. The temple's inscriptions offer valuable insights into the scientific and literary advancements of the era.

George Michell affirms:

Temples in India were not merely places of worship but also centres of learning and cultural activity. They housed libraries, schools, and workshops where scholars, artists, and craftsmen gathered to study and create. The inscriptions on temple walls served as public records, documenting historical events, mathematical calculations, and astronomical data. Temples like the Brihadeeswarar Temple in Thanjavur and the Sun Temple at Konark were not just architectural marvels but also repositories of knowledge, reflecting the integration of science, art, and spirituality in Indian culture. (67)

The Jagannath Temple in Puri, which served as a prominent educational centre for the study of Vaishnavism, Sanskrit literature, and Odissi dance. The temple complex facilitated theological debates, literary compositions, and the transmission of spiritual and artistic traditions, underscoring its role as a dynamic institution of knowledge and culture. These examples illustrate how temples transcended their religious functions to become vital centres of intellectual and cultural advancement in medieval India.

Historical Context

The colonial period marked a profound transformation in India's education system, systematically replacing indigenous institutions like *gurukuls* and *pathshalas* with Western-style schooling. This shift deliberately marginalized traditional knowledge in fields such as medicine, astronomy, mathematics, and philosophy, dismissing them as outdated while promoting Western paradigms as superior. The disruption disproportionately affected women, who were primary transmitters of traditional knowledge, as colonial policies emphasized male-dominated institutions, side lining women's roles.

Post-independence, India's education system continued to prioritize Western frameworks, deepening the marginalization of indigenous practices. The focus on Western science, technology, and humanities left little room for traditional knowledge, eroding cultural identity and heritage among younger generations. This systemic marginalization has hindered the preservation of India's intellectual heritage and limited opportunities for holistic, culturally rooted learning. Addressing this historical disconnect requires integrating indigenous wisdom into modern education, ensuring future generations benefit from India's rich cultural and intellectual legacy. Basham A. L. states, "The temples of ancient India were not merely places of worship but also centres of learning, where scholars gathered to debate and disseminate knowledge on subjects ranging from metaphysics to mathematics". (74)

The Need for Decolonization

The integration of Indian Knowledge Systems (IKS) into contemporary education is essential for fostering cultural identity and pride among students. For decades, Western-centric curricula have created a disconnect between students and their heritage. Incorporating IKS helps students reconnect with their roots, instilling pride in India's rich intellectual traditions, such as contributions to mathematics, astronomy, and philosophy. This connection is vital in a globalized world where cultural identity is at risk of erosion. Moreover, IKS promotes holistic, interdisciplinary learning, combining science, philosophy, arts, and ethics, aligning with the National Education Policy (NEP) 2020's vision of well-rounded education. Systems like Ayurveda and Yoga integrate physical, mental, and spiritual well-being, while classical arts and literature foster creativity and critical thinking. Embracing IKS equips students with the skills to navigate complex challenges, moving beyond rote learning to a more comprehensive educational framework.

Another compelling reason for decolonizing education is the potential of Indian Knowledge Systems

(IKS) to contribute to sustainable development. Traditional practices in agriculture, water management, and health, such as stepwells, rainwater harvesting, and organic farming, offer time-tested solutions to modern environmental and social challenges. Integrating these practices into education equips students to apply indigenous wisdom for sustainable and equitable solutions. Dr. A. P. J. Abdul Kalam affirms:

Ancient India was a knowledge society that contributed immensely to the world in the fields of mathematics, astronomy, medicine, and philosophy. The concept of zero, the decimal system, and the value of pi were all gifts from India to the world. The Vedas and Upanishads are not just religious texts but also repositories of scientific thought and philosophical inquiry. India's ancient universities, such as Nalanda and Takshashila, were centers of learning that attracted scholars from across the globe. (67)

Furthermore, IKS holds global relevance, with concepts like *Dharma* (duty) and *Karma* (action) providing ethical frameworks that resonate across cultures, and contributions like the concept of zero impacting global knowledge systems. By incorporating IKS, India can contribute to global dialogues on sustainability, ethics, and innovation. Radhakrishnan S. states, "The philosophical concepts of Dharma and Karma in the Indian Knowledge System provide timeless ethical frameworks that guide individual and societal conduct, transcending cultural and temporal boundaries" (38). Thus, decolonizing education through IKS integration reclaims cultural heritage while fostering holistic, sustainable, and globally relevant learning, empowering students to connect with their roots, address contemporary challenges, and contribute meaningfully to the global community.

Strategies for Integrating Indian Knowledge Systems (IKS)

A key strategy for integrating Indian Knowledge Systems (IKS) into mainstream education is curriculum reform, introducing IKS at all educational levels through an interdisciplinary approach that connects it with subjects like science, mathematics, medicine, and environmental studies. Another crucial step is research and documentation, as much of India's traditional knowledge has been orally transmitted. Systematic academic research is needed to document, analyse, and validate IKS, making it accessible and credible within modern frameworks. These efforts can bridge the gap between traditional and contemporary knowledge, fostering greater acceptance and relevance of IKS in education.

Equally important is teacher training to equip educators with the skills to effectively teach IKS. Specialized programmes should familiarize teachers with its historical, scientific, and philosophical foundations, enabling seamless integration with modern education. Community involvement is also essential, with local scholars and practitioners contributing to curriculum development and teaching, ensuring authenticity and depth. Additionally, digital platforms can transform IKS accessibility through online courses, video lectures, and interactive databases. Leveraging technology to create digital repositories of ancient texts and research findings can broaden engagement, ensuring IKS remains relevant in the digital age. By implementing these strategies, IKS can be integrated into mainstream education, preserving its legacy while addressing contemporary challenges. Vandana Shiva opines:

India's traditional knowledge systems, especially in agriculture, are based on the principles of sustainability, diversity, and respect for nature. The Green Revolution, with its focus on monocultures and chemical inputs, disrupted these systems, but the wisdom of our ancestors still holds the key to solving the ecological crises we face today. Traditional practices like organic farming, seed saving, and water harvesting are not just alternatives; they are the foundation of a sustainable future. (89)

Case Studies on the Integration of Indian Knowledge Systems (IKS)

A significant example of integrating Indian Knowledge Systems (IKS) into modern education is the inclusion of Ayurveda in medical curricula, promoting a holistic approach to healthcare. Ayurveda's focus on preventive medicine, personalized treatment, and natural remedies complements modern biomedical practices, enabling healthcare professionals to offer comprehensive treatment options. Similarly, Indian mathematics, particularly Vedic mathematics, is being incorporated into STEM education for its efficiency in calculations. Contributions of ancient scholars like Aryabhata, Brahmagupta, and Bhaskara in algebra, trigonometry, and calculus are also recognized, enriching students' mathematical skills and problem-solving abilities while fostering appreciation for India's intellectual heritage. B. V. Subbarayappa, asserts: "Ancient Indian sciences, including astronomy, mathematics, and medicine, were deeply rooted in empirical observation and logical reasoning, challenging the notion that traditional knowledge lacks scientific rigor" (45).

In the field of **environmental studies**, traditional ecological knowledge is being integrated into curricula to address contemporary environmental challenges. Indigenous water conservation development and teaching, ensuring authenticity and depth. Additionally, digital platforms can transform IKS accessibility through online courses, video lectures, and interactive databases. Leveraging technology to create digital repositories of ancient texts and research findings can broaden engagement, ensuring IKS remains relevant in the digital age. By implementing these strategies, IKS can be integrated into mainstream education, preserving its legacy while addressing contemporary challenges. Vandana Shiva opines:

India's traditional knowledge systems, especially in agriculture, are based on the principles of sustainability, diversity, and respect for nature. The Green Revolution, with its focus on monocultures and chemical inputs, disrupted these systems, but the wisdom of our ancestors still holds the key to solving the ecological crises we face today. Traditional practices like organic farming, seed saving, and water harvesting are not just alternatives; they are the foundation of a sustainable future. (89)

Case Studies on the Integration of Indian Knowledge Systems (IKS)

A significant example of integrating Indian Knowledge Systems (IKS) into modern education is the inclusion of Ayurveda in medical curricula, promoting a holistic approach to healthcare. Ayurveda's focus on preventive medicine, personalized treatment, and natural remedies complements modern biomedical practices, enabling healthcare professionals to offer comprehensive treatment options. Similarly, Indian mathematics, particularly Vedic mathematics, is being incorporated into STEM education for its efficiency in calculations. Contributions of ancient scholars like Aryabhata, Brahmagupta, and Bhaskara in algebra, trigonometry, and calculus are also recognized, enriching students' mathematical skills and problem-solving abilities while fostering appreciation for India's intellectual heritage. B. V. Subbarayappa, asserts: "Ancient Indian sciences, including astronomy, mathematics, and medicine, were deeply rooted in empirical observation and logical reasoning, challenging the notion that traditional knowledge lacks scientific rigor" (45).

In the field of **environmental studies**, traditional ecological knowledge is being integrated into curricula to address contemporary environmental challenges. Indigenous water conservation techniques, such as step-wells, tank irrigation, and rainwater harvesting, have been revived in sustainable development programs. Similarly, traditional agricultural practices like organic farming, crop rotation, and agroforestry offer valuable insights into sustainable land management. By incorporating these traditional ecological practices into education, students can learn environment-friendly solutions rooted in India's long history of ecological wisdom. These case studies highlight the potential of integrating IKS into modern education to enhance learning outcomes and promote sustainable, holistic approaches to knowledge. By embracing such initiatives, educational institutions can bridge the gap between tradition and modernity while ensuring the continued relevance of India's vast intellectual traditions. Mahatma Gandhi aptly asserts: "The Indian knowledge system is holistic, encompassing the physical, mental, and spiritual dimensions of life". (33)

Conclusion

Temples in medieval India were dynamic centres of knowledge and learning, nurturing scholars, advancing scientific inquiry, and preserving artistic traditions. Recognizing this legacy offers valuable lessons for modern education, as the holistic, interdisciplinary approach of temple-based learning can inspire contemporary pedagogy, fostering a deeper connection with India's heritage. Decolonizing Indian education by integrating Indian Knowledge Systems (IKS) into mainstream curricula is both a cultural imperative and a necessity for holistic, sustainable development. The exclusion of indigenous knowledge has created a disconnect between India's intellectual heritage and its modern educational framework. Reclaiming this heritage empowers future generations to think critically, act sustainably, and take pride in their cultural identity. By bridging traditional and contemporary knowledge, the education system can become more inclusive and equitable, honouring the past while preparing students for future challenges. Integrating IKS is a transformative step toward a globally relevant, culturally rooted education system.

Works Cited

- Basham, A.L. *The Wonder That Was India*. Grove Press, 1954.
- Coomaraswamy, Ananda K. *The Dance of Shiva: Essays on Indian Art and Culture*. Dover Publications, 1985.
- Gandhi, Mahatma. *Hind Swaraj or Indian Home Rule*. Navajivan Publishing House, 1938.
- Goswamy, B. N. *The Essence of Indian Art*. Penguin, 1986.
- Kalam, A.P.J. Abdul, and Y.S. Rajan. *India 2020: A Vision for the New Millennium*. Penguin Books, 1998.
- Kramrisch, Stella. *The Hindu Temple*. Motilal Banarsidass, 1976.
- Michell, George. *The Hindu Temple: An Introduction to Its Meaning and Forms*. University of Chicago Press, 1988.
- Radhakrishnan, S. *The Hindu View of Life*. HarperCollins, 1993.

- ---. *Indian Philosophy*. Oxford University Press, 1923.
- Rao, Balachandra. *Indian Astronomy: An Introduction*. OUP, 2000.
- Shiva, Vandana. *Soil Not Oil: Environmental Justice in an Age of Climate Crisis*. South End Press, 2008.
- Subbarayappa, B.V. *The Tradition of Astronomy in India*. Centre for Studies in Civilizations, 2008.
- Tagore, Rabindranath. *Towards Universal Man*. Asia Publishing House, 1961.

Unveiling the Feminine Voice of Gargi Vachaknavi in the Realm of Vedic Philosophy

Radhika R

Assistant Professor in English, HHMSPB NSS College for Women, Neeramankara, Thiruvananthapuram

Abstract

Women philosophers played a significant, though often overlooked, role in the development of Vedic literature and philosophy. Women such as Gargi Vachaknavi, Maitreyi, and others defied societal norms and emerged as profound intellectual figures, contributing to philosophical discourses and shaping the spiritual and intellectual framework of Vedic thought. This paper delves into Gargi's philosophical legacy, especially exploring her role in the dialogues of the Brihadaranyaka Upanishad, where she boldly engaged with male sages on matters of metaphysics, cosmology, and the nature of the self. It also sheds light on the broader implications of Gargi's role in Vedic thought, not just as a sage, but as a symbol of the feminine voice in the traditionally patriarchal realm of Vedic philosophy. Her insight into the nature of reality, the self, and existence have left an indelible mark on the development of Vedic philosophy. Through her engagement, Gargi's voice continues to resonate, advocating for the intellectual empowerment of women in ancient India and beyond.

Keywords: Vedic Philosophy,Brihadaranyaka Upanishad, *Brahman, Atman,* Rigveda

Indian philosophy is a vast and diverse intellectual tradition that encompasses a wide range of ideas, schools, and systems of thought. It has evolved over thousands of years and has deeply influenced the spiritual, ethical, metaphysical, and practical aspects of Indian culture. Indian philosophy is known for its emphasis on questions of existence, the nature of reality, consciousness, and the ultimate purpose of life. Vedic philosophy holds a foundational and crucial role in the development of Indian philosophy. The Vedas, which are the oldest sacred texts of India, form the bedrock of many of the central ideas, principles, and metaphysical inquiries that later flourished in the various schools of Indian philosophical thought. Vedic philosophy not only set the stage for the philosophical systems that followed but also deeply influenced the ethical, spiritual, and intellectual life of India for thousands of years.

In the ancient tapestry of Indian philosophy, the voices of many philosophers have been woven into the fabric of time, shaping the contours of spiritual and intellectual thought. The Vedic tradition, is often regarded as a male-dominated intellectual tradition. However, women have played a significant, though often overlooked, role in the development and evolution of Vedic thought. In the ancient period, women philosophers, poets, and sages contributed meaningfully to the intellectual and spiritual discourse, especially within the texts of the *Vedas,* the *Upanishads,* and the *Brahmanas.* These women's voices were powerful and enduring, and they were integral to the philosophical debates and theologies that helped shape the trajectory of Vedic philosophy. Although the narratives of these women are not as extensively documented as those of their male counterparts, their contributions are crucial to understanding the full scope of Vedic thought and its emphasis on spiritual wisdom, knowledge, and the metaphysical realities of the universe. Figures like Gargi Vachaknavi, Maitreyi, Apala, and others demonstrate that women were not passive figures in the Vedic tradition but active contributors to philosophical discourse. Their philosophical inquiries on the nature of existence, knowledge, and the cosmos reflect a deep and insightful engagement with Vedic thought.

Among these towering women figures, Gargi Vachaknavi stands as one of the most remarkable philosopher, poet, and sage. She was an extraordinary woman from ancient India who lived from the 9th to the 7th centuries BCE. Born to the sage Vachaknu, she came from a family known for its deep wisdom and spirituality. From an young age itself, Gargi was incredibly curious about Vedic scriptures and philosophy, mastering complex ideas that many men of her time struggled with. The unique thing about Gargi was, her choice to stay unmarried, which was a daring decision especially at her times, eventhough her mother wanted her to get married. Most probably this independence might be one of the reasons to let her to focus entirely on her quest for knowledge. She was known as Brahmavadini, a woman who seeks for the knowledge of the ultimate reality, Brahman, which is considered as the highest form of knowledge that leads to spiritual liberation or moksha.She was born at a time when the Vedic texts were being compiled, and the early Upanishads were being composed. It was a period marked by deep metaphysical exploration and the search for ultimate truths about existence, the universe, and the nature of the soul. Her life and works provide insights not only into the philosophical landscape of ancient India but also into the role of women in intellectual and spiritual traditions.

Gargi Vachaknavi holds a prominent and revered position in the Indian knowledge system, particularly within the context of ancient Indian philosophy and spiritual thought. She had mastered all four Vedas and was constantly engaged in discussions and debates that highlighted her grasp of deep spiritual concepts. She is often associated with the Upanishadic teachings, particularly in the *Brihadaranyaka Upanishad*, one of the oldest Upanishads in the corpus of Vedic texts. The **Brihadaranyaka Upanishad** is one of the oldest and most significant of the Upanishads, which are the concluding portions of the Vedic texts. It is considered a key philosophical text in the Indian spiritual tradition, and it is part of the **Yajurveda**. The name "Brihadaranyaka" can be translated as "the great wilderness" or "the great forest," symbolizing the vast and profound nature of the knowledge contained within it. In this text, Yajnavalkya plays a central role as both a teacher and a debater, imparting profound spiritual wisdom. The setting for the dialogues is a grand philosophical assembly held at the Rajasuya Yajna, King Janaka's court of Videha Kingdom, where sages, philosophers, and intellectuals from various parts of ancient India had gathered. The central theme of the assembly was to discuss profound questions concerning the nature of reality, the soul, and the ultimate truth. He is known for his deep insights into the nature of *Brahman* (the ultimate reality) and *Atman* (the individual soul). His famous declaration, "Tat Tvam Asi"

("That Thou Art"), is a cornerstone of Vedantic philosophy, signifying the realization that the individual soul (Atman) is identical with the supreme universal consciousness (Brahman). This teaching emphasizes the non-duality (Advaita) of the individual and the universal. In the *Brihadaranyaka Upanishad*, Yajnavalkya is depicted as a sage who has reached the highest state of spiritual knowledge and wisdom, and he advocates for renouncing the ego and material possessions in order to realize one's true nature. One of the most famous sections of the Brihadaranyaka Upanishad is the dialogue between the sage Yajnavalkya and his wife Maitreyi. In their conversation, Yajnavalkya explains the nature of immortality and the ultimate reality to Maitreyi, demonstrating the importance of understanding Brahman in order to attain liberation. Maitreyi asks her husband whether wealth or material possessions can lead to immortality. Yajnavalkya responds by explaining that only the knowledge of Brahman can lead to liberation, and that wealth cannot provide immortality. This dialogue reflects the importance of spiritual wisdom over worldly possessions.

Another notable exchange of dialogue is between **Yajnavalkya** and **Gargi Vachaknavi.** She engages in a series of philosophical dialogues with the great sage Yajnavalkya. Gargi Vachaknavi, challenges him with questions that probe the very essence of existence. It also delves into the complex metaphysical topics such as the nature of the cosmos, the eternal and infinite nature of *Brahman,* and the concept of *Atman* (the individual soul).

The central theme of the dialogue between Gargi and Yajnavalkya revolves around the nature of *Brahman,* the formless, infinite, and all-encompassing ultimate reality in Vedic thought. In the words of Yajnavalkya, Brahman is that from which all beings are born, that by which, when born, they live, and that into which, when they die, they enter. It is the **sustaining power** that animates the universe and gives life to all living beings. Brahman is the final **destination** of all things, the **ultimate reality** into which everything merges at the end of its cycle. Gargi, known for her intellectual acumen, questions Yajnavalkya about the true nature of this supreme reality. One of the most famous exchange in the *Brihadaranyaka Upanishad* is the question asked by **Uddalaka Aruni**, a student of the sage**Yajnavalkyathat, what pervades everything and remains after everything else dissolves to which**Yajnavalkya explains that it is *Brahman* that transcends all material forms and remains after the dissolution of the universe, **which again is** connected to the philosophical inquiries that **Gargi Vachaknavi** engages in during her dialogues in the same text. This dialogue reflects Gargi's understanding of the metaphysical aspects of the cosmos and her willingness to ask challenging questions even in the presence of great sages like Yajnavalkya. Gargi's questions touch upon the infinite and formless nature of *Brahman*, highlighting her recognition of the importance of this concept in the spiritual and philosophical understanding of existence.

Another key aspect of the dialogue concerns the concept of *Atman* or the individual soul. For **Sage Yajnavalkya, Atman** (the Self) is the **eternal, unchanging, and indestructible essence** that resides within every living being. It is the true nature of the individual, beyond the physical body, mind, and ego. Yajnavalkya expounds on the concept of **Atman** in several key sections of the *Brihadaranyaka Upanishad*. The **Atman** is not separate from **Brahman** (the ultimate reality). The individual **Atman** is a reflection of the **Brahman**, and realization of this truth leads to self-realization and liberation (moksha). Yajnavalkya explains that Atman is **indestructible, eternal,** and**imperishable**. The physical body may perish, but the Atman within remains unaffected by the cycles of birth and death. Gargi questions the nature of the relationship between *Atman* (the individual soul) and *Brahman*(the ultimate reality). This exploration reflects the Upanishadic philosophy, which seeks to understand the union of the individual soul with the universal self. The realization of **Atman** leads to liberation from the cycle of **samsara** (birth, death, and rebirth). Once an individual recognizes that their true identity is not the body or mind, but the **Atman**, they achieve liberation (moksha) by understanding their unity with Brahman. Gargi's inquiry into the nature of *Atman* and its connection to the universe demonstrates her understanding of the central themes of Vedic and Upanishadic thought: the search for the true self and the realization of its oneness with *Brahman*. This dialogue contributes significantly to the philosophical discussions on the nature of existence, the soul, and ultimate reality.

In her exchange with Yajnavalkya, Gargi also delves into the concept of transcendence and immanence. She asks about the ultimate source of all creation, exploring how the material world and the spiritual world are interconnected. Gargi's questions challenge the conventional understanding of the universe, suggesting that *Brahman* is both transcendent (beyond the physical world) and immanent (pervading all things). This reflects the non-dualistic approach of the Upanishads, where the divine is seen as transcendent and immanent, existing within and beyond the material realm.

The final part of the dialogue between Gargi and Yajnavalkya is marked by Gargi's question on the ultimate source of all existence. As the conversation deepens,

Gargi begins to realize the vastness of the mystery she is grappling with, and she finally concludes her line of questioning by asking a final question about the nature of the ultimate reality. Yajnavalkya responds, explaining the ineffable nature of *Brahman*, indicating that it cannot be comprehended by ordinary human understanding. Gargi, acknowledging the limits of human knowledge, humbly accepts that the ultimate truth is beyond human comprehension. This marks the end of her questioning in the dialogue. The conclusion of this exchange emphasizes the limits of human knowledge and the ineffable nature of the ultimate reality, a theme that is central to many of the Upanishadic texts. The most debated question asked by Gargi with Yajnavalkya is, 'since this whole world is woven back and forth on water, on what then is woven back and forth' proves her mastery on the cosmology. It is even stated that if Yajnavalkya can answer the questions posed by Gargi, no one else can defeat him in the philosophical thoughts. That itself proves Gargi's indepth knowledge regarding her intellectual mastery on the vedic thoughts. She was even recognized as one of the Navaratnas(Nine Jewels) in the court of King Janak of Mithila.

Gargi's engagement with Yajnavalkya reveals her as an intellectually formidable figure that was deeply immersed in the metaphysical and philosophical inquiries of her time. In the dialogues, she is not portrayed as a passive listener, but as an active participant in the pursuit of knowledge. Her questions challenge not only Yajnavalkya but also the intellectual status quo, highlighting the fact that women were not excluded from such high-level philosophical discussions in ancient India. Her exemplary question about what lies beneath the earth and what holds it all together dig deeper into the nature of existence. Gargi's role in these dialogues is emblematic of the Vedic tradition's openness to intellectual inquiry and philosophical debate, irrespective of gender. Her courage to ask difficult questions and challenge the ideas of the most revered sages of her time reflects her intellectual independence and philosophical depth.

Gargi's inquiries are remarkable not only because of their intellectual depth but also because they were directed toward a male sage in an open forum. At a time when women were largely excluded from such intellectual exchanges, Gargi's participation reflects the progressive nature of Vedic philosophy, where knowledge and wisdom were valued over gender. Through her dialogues, Gargi also embodies the ideal of *Jnana* (knowledge) as something that is accessible to all, regardless of gender, reinforcing the universal nature of Vedic wisdom. By engaging in dialogues on the nature of the universe and human existence, Gargi carved out a space for women philosophers in the ancient tradition, laying the groundwork for the inclusion of feminine perspectives in subsequent philosophical discussions. Her work emphasizes that, the essence of Vedic philosophy is not confined to a single viewpoint but is enriched by multiple perspectives—whether they be masculine or feminine.

Gargi was also well-versed in the hymns of the *Rigveda* and was recognized for her profound understanding of Vedic knowledge and cosmology. It is considered that she even wrote many hymns in Rigveda. The *Rigveda* speaks of **Purusha,** the cosmic being, whose sacrifice leads to the creation of the universe. Gargi's philosophical questions explore these themes and seek to understand the nature of reality at a fundamental level, much like the philosophical inquiries expressed in the Vedic hymns themselves. As a scholar of the Vedas, Gargi would have been intimately familiar with the hymns of the *Rigveda*, which form the oldest part of the Vedic corpus. These hymns primarily consist of prayers, invocations, and philosophical reflections on the nature of the universe, the gods, and the cosmic order (Rita). Her name is even mentioned in *Grihya Sutras of Asvalayana*, which delves into the details of domestic religious rituals and ceremonies performed by householders consisting of different stages of life from conception to death. Her contributions can also be seen in the text *The Yoga Yajnavalkya*, which is based on the dialogue between Gargi and Yajnavalkya on yoga. The term 'I am a soul' was given by her and is mentioned in the vedas. She was also very popular during the Treta Yug for mentoring the four Mithila sisters, taking responsibility of their education & upbringing them according to their future life. Gargi's voice continues to resonate, not only as a symbol of intellectual achievement but also as a beacon for inclusivity in the intellectual traditions of the world with women playing a critical role in the shaping of the metaphysical and spiritual landscape of ancient India. No wonder Gargi Vachaknavi is regarded as the first women philosopher in ancient India especially in the Vedic tradition.

References

- Iyengar, S. Ramaswamy. *Women in the Vedic Period.* Bharatiya Vidya Bhavan, 1995.
- Khanna, B. R. *Philosophy and Women in Ancient India.* Motilal Banarsidass, 2006.
- O'Flaherty, Wendy Doniger. *The Upanishads: A New Translation.* Penguin Books, 2008.
- Bandyopadhyay, J. *Women in Ancient Indian Society.* Vikas Publishing House, 2001.
- Raju, P.T. *The Philosophical Traditions of India.* Oxford University Press, 2007.

Sustainability and Environmental Economics in Indian Traditions: Bridging Ancient Wisdom with Modern Policy

Dr. Sangeetha U V

Assistant Professor, Department of Economics, VTM NSS College Dhanuvachapuram, Thiruvananthapuram, Kerala, India, MOBILE- 7736352306, email- sangeethandd86@gmail.com

ABSTRACT

For thousands of years, Indian civilisation has been rooted on sustainability, which is reflected in its economic, religious, and philosophical traditions. Ancient Indian sustainability was firmly anchored in ecological, ethical, and spiritual balance, in contrast to the contemporary idea, which frequently emphasises technology solutions. The interdependence of all living forms was highlighted in the Vedas and Upanishads, which promoted the prudent use of natural resources. The necessity of harmony between humans and environment is emphasised by the Rig Vedic notion of RTA (cosmic order).

Keywords: Ancient knowledge, Indian tradition, Sustainability, Arthashastra, Vedhas

INTRODUCTION

Sustainability is meeting the needs of the present without compromising the ability of future generations to meet their own needs. Environmental economics is a branch a of economics that studies the economic impact of environmental policies and the efficient use of natural resources. Indian traditions have long emphasized harmony between nature and economic activities. Sustainable economic concepts are emphasised in ancient literature such as the Vedas, Arthashastra, Upanishads, and Buddhist and Jain teachings. The Atharva Veda advocates for sustainable land and water management, referring to the Earth as Bhumi Mata, or Mother Earth. Water management, pollution prevention, and forest protection were all covered in depth in Kautilya's Arthashastra. Mindful consumerism and simplicity were encouraged by Buddhism's Middle Path. Ahimsa, or non-violence, is a principle of Jainism that discourages excessive resource usage and deforestation. An early example of environmental legislation was found in Manusmrithi, which established punishments for damaging the environment. Dharma (responsibility) in environmental ethics includes obligations to protect the environment.

Statement of the Problem

Unsustainable economic practices and industrialisation have led to contemporary environmental problems such pollution, water shortages, deforestation, and climate change. Modern sustainability initiatives sometimes ignore old knowledge systems that have successfully preserved ecological balance for millennia, even if they mostly rely on technology advancements and policy-driven strategies. A wealth of environmental ethics and sustainable economic ideas may be found in ancient Indian traditions and scriptures, such as the Vedas, Upanishads, Arthashastra, and Buddhist and Jain writings. However, contemporary environmental policies have mostly ignored these tried-and-true techniques.

SIGNIFICANCE OF THE STUDY

Modern environmental policies may learn a lot from the ancient Indian approach to sustainability, which blended ecological protection with economic well-being. The problems of resource depletion, climate change, and sustainable development can be addressed by combining traditional knowledge with modern science.

OBJECTIVES OF THE STUDY

1. To explore how traditional Indian knowledge promoted environmental conservation
2. To discuss how ancient principles can be applied to modern sustainability challenges

SCOPE OF THE STUDY

The purpose of this study is to investigate how traditional Indian knowledge supported sustainable resource management and environmental preservation. In order to solve contemporary sustainability issues, it seeks to investigate the applicability, relevance, and potential integration of traditional Indian environmental ideas. In order to close the gap between historical knowledge and contemporary sustainability practices, the study will evaluate if these antiquated methods may strengthen or supplement current environmental measures.

Limitations of the study

1. Lack of direct empirical data, as the study relies solely on secondary sources.
2. Possible gaps in historical records, making interpretation subjective
3. Translation challenges for ancient texts that may have different contextual meaning today.

REVIEW OF LITERATURE

Scholars such as Sharma (2015) and Bhardwaj (2018) argue that ancient Indian scriptures, including the *Vedas*, *Upanishads*, and *Puranas*, emphasize a symbiotic relationship between humans and nature. The concept of *Dharma* (duty) in Hindu philosophy includes environmental ethics, advocating for the preservation of natural resources. The *Atharva Veda* explicitly mentions conservation practices, promoting a holistic view of nature as sacred (Mukherjee, 2020).. Studies by Kumar (2017) and Patel (2019) highlight how these teachings encourage minimal exploitation of natural resources, aligning with contemporary sustainability goals.Traditional Indian agricultural practices have been examined by Prasad (2016) and Desai (2021), who emphasize the role of organic farming, crop rotation, and water conservation techniques such as *Bamboo Drip Irrigation* and *Tank Irrigation Systems*. Furthermore, the concept of *Chakra Vyuh* in economic planning, as analyzed by Singh (2022), demonstrates how ancient strategies focused on a circular economy, minimizing waste and optimizing resource allocation.

Additionally, festivals such as *Tulsi Vivah* and *Nag Panchami* celebrate nature, reinforcing environmental values among communities. A comparative study by Chatterjee (2021) examines how such traditions align with contemporary sustainability principles like biodiversity conservation and ethical consumerism.

RESEARCH GAP

Despite extensive research on India's traditional environmental wisdom, gaps remain in empirical studies measuring its direct impact on contemporary policy outcomes. Future research could focus on quantifying the effectiveness of traditional conservation techniques in mitigating climate change and their adaptability in urban sustainability models.

METHODOLOGY OF THE STUDY

The data used in this investigation is secondary. The research employs a descriptive, analytical, and qualitative methodology. In order to comprehend traditional sustainability ideas and their applicability today, the research will examine current literature, historical texts, policy papers, and case studies.

ANALYSIS
Objective 1
To explore how traditional Indian knowledge promoted environmental conservation

This study examines the ways in which ancient Indian knowledge systems supported sustainability and contrasted them with contemporary strategies for environmental preservation. Numerous traditional methods for conserving water, managing forests, farming, and reducing waste are being resurrected and incorporated into current environmental policy. Traditional Indian knowledge systems provide long-term, sustainable answers to contemporary environmental problems. Here are a few significant ways that traditional Indian knowledge is currently supporting environmental preservation.

Table 1.1: Ancient Water Management and Modern Water Conservation

Aspects	Traditional Indian Knowledge	Modern Application of Ancient Sustainability Practices
Water Harvesting	Stepwells (baoils), Tank Irrigation, Johads(rain water harvesting structure Temple Tanks and Sacred Water bodies Community -led water management (Phad irrigation system in Maharastra, Eri system in Tamil Nadu) managed local water resources efficiently	Dams, Reservoirs, Rainwater harvesting systems Jai Shakthi Abhiyan and Mission Amrit Sarovar are reviving ancient water conservation techniques. Community Based Water Management Programme are using traditional wisdom for rural water security
River Conservation	Rivers considered sacred (Ganga, Yamuna) with rituals ensuring their protection	National River Conservation Plan (Namami Gange) River rejuvenation projects

Source: Secondary Source

Interpretation

In order to ensure sustainable usage and long-term groundwater recharge, traditional methods (such as stepwells and temple tanks) concentrated on community-based and decentralised water management. Huge-scale infrastructure projects like dams and reservoirs are examples of modern techniques.These projects offer water security, but they can also cause ecological harm (such as relocation from huge dam projects). Rivers' sacred importance in traditional systems prompted conservation efforts, but industrial pollution has made contemporary river rejuvenation initiatives necessary. For sustainable water usage, the best strategy combines contemporary microirrigation methods with conventional rainwater gathering.

Table 1.2 Organic and Sustainable Agriculture inspired by Ancient Farming System

Aspects	Traditional Indian Knowledge	Modern application of ancient sustainability practices
Farming Techniques	Organic farming, mixed cropping crop rotation, to maintain soil fertility (described in Vedic texts)	Intensive farming, Monocropping, Zero Budget Natural Farming (ZBNF) and Permaculture
Pest Control	Natural pesticides (Neem, cow-dung,turmeric, buttermilk)	Chemical pesticides (causing soil degradation)
Soil Conservation	Use of natural manure, Agroforestry systems integrating tree cultivation with crop	Synthetic fertilizers, land reclamation projects

Source: Secondary Source

Interpretation

Crop rotation, intercropping, and organic fertilisation were among the natural techniques utilised in traditional agriculture to maintain soil fertility and ecosystem health. High productivity is a top priority in modern agriculture, yet overuse of fertilisers and pesticides frequently results in soil deterioration, chemical pollution, and biodiversity loss. Organic farming is supported by the government through the Paramparagat Krishi Vikas Yojana using ancestral crops and traditional seeds to promote resilient farming.

Table 1.3 Biodiversity and Forest Conservation

Aspects	Traditional Indian Knowledge	Modern application of ancient sustainability practices
Forest Protection	Sacred Groves (Devrai, Devarakadu, Sama, Kavu) Community managed forests	Sacred groves are now recognized as Biodiversity Hotspots and protected y government initiatives in Kerala, Karnataka and Rajasthan Reserved forests, Wildlife sanctuaries, National parks
Wild life Protection	Ahimsa, protection of keystone species in religious ous texts	Conservation laws (Wildlife Protection Act,1972), Eco tourism
Afforestation	Vedic rituals promoting tree planting (peepal, banyan, tulsi)	Afforestation (Green India Mission) and Reforestation programs

Source: Secondary Source

Interpretation

Sacred groves were an early example of community-driven biodiversity protection that helped to preserve and protect forests. Although urbanisation and industrialisation continue to destroy habitat, modern afforestation initiatives like Green India Mission are working to recover destroyed forests.

Tale 1.4: Circular Economy and Zero Waste Principles

Aspect	Traditional Indian Knowledge	Modern application of ancient sustainability practices
Waste disposal	Minimal waste culture (reuse of materials, composting traditional clay utensils)	Waste management systems, Recycling plants Plastic Free Movements inspired by traditional zero waste practices (banning plastic plates, encouraging banana leaf use) Government Swatch Bharath Mission encourages waste management using traditional techniques.
Circular economy	Temple economies (eg: reusing flowers for incense, prasadam distribution)	Temple waste recycling programs convert floral waste into compost and incenses Corporate circular economy models

Source: Secondary Source

Interpretation

Waste was reduced and repurposed in traditional cultures' circular economies (e.g., temple offering transformed into compost). Industrial waste and plastic pollution are still big issues even though modern waste management depends on technology solutions (such as recycling plants).

Table1.5: Ethical Business and Sustainable Economic Policies

Aspects	Ancient Indian Knowledge	Modern applications of ancient sustainability practices
Governance model	Village based environmental management, local water and forest councils (Gramsahas, Panchayats)	Centralized environmental policies, Government funded programmes
Pollution control	Ethical duty (dharma) to protect the environment, punishment to polluters in Arthashastra	Environmental laws (Air Act 1981, Water Act1974, National Green Tribunal 2010)
Sustainability policies	Emphasis on local self-sufficiency (Swadeshi)	Sustainabale Development Goals (SDGs), Environmental, Social Governance(. ESG) policie, Government Athmanirbhar Bharat and Make in India programs promote sustainable .self reliant productions
		Carbon Credit and Green Finance policies reflect ancient responsibility driven economies

Source: Secondary Source

Interpretation

Community-led environmental management, such as village councils controlling water usage and deforestation, was a key component of ancient government regimes. Modern government is based on the law, but because of political obstacles and corporate interests, enforcement is frequently weak.

Objective 2: To discuss how ancient principles can be applied to modern sustainability challenges

Though based on spiritual, cultural, and philosophical beliefs, ancient Indian knowledge systems provide profound insights into ethical governance, sustainable living, and environmental stewardship. These ideas can be applied to contemporary sustainability issues to support sustainable development and environmental preservation. The following lists important traditional Indian ideas and how they are being used now to solve urgent sustainability issues.

Table 1.6: Application of ancient principles into modern sustainability challenges

Ancient principle	Modern application
Concept of Ahimsa (Non-Violence) is one of the central tenets of Indian philosophy in Jainism, Buddhism and Hinduism. It extends beyond human interactions to include non-harm towards all living beings, including animals, plants and the environment	Environmental Ethics and Animal Rights: Ahimsa can drive modern ethical consumerism emphasizing cruelty-free and sustainable products. The principle is applied in animal protection laws, biodiversity conservation and ethical treatment of nature. Conservation of Resources: In farming the principle encourages natural, chemical-free agriculture, reducing harm to soil health, water bodies and wild life through the use of organic farming practices and pesticide reduction

Sacred Groves and Biodiversity Protection Sacred Groves (Devrai, Devarakadu, Kavu) were patches of forest or trees deemed sacred and protected by local communities. These areas, often located near temples or in natural sites, were preserved for spiritual and ecological reasons.	Biodiversity Hotspots: Sacred groves can be seen as community-driven biodiversity conservation areas, where local people voluntarily protect flora and fauna. Modern conservation efforts like wildlife sanctuaries and community-managed protected areas can adopt similar models. Forest Restoration and Afforestation: Sacred groves' principles of sustainable forest management can inform afforestation programs and the restoration of degraded ecosystems by emphasizing cultural value and community participation.
Vasthu Shastra and Sustainable Architecture Vastu Shastra, the ancient Indian architectural philosophy, promotes the design of spaces that are harmonious with nature. It includes principles like natural lighting, ventilation, energy-efficient materials, and alignment with the elements (earth, water, fire, air, space).	Green Building and Architecture: Modern buildings can incorporate Vastu principles for energy efficiency, solar passive designs, and natural resource optimization. Elements such as rainwater harvesting, natural cooling, and solar energy utilization can reduce the carbon footprint of urban developments. Urban Sustainability: In dense urban areas, Vastu-inspired architecture can improve the sustainability of living spaces by ensuring optimal use of natural resources and energy conservation.

Dharma (Ethical Responsibility) and Corporate Sustainability Dharma refers to the moral order and the duties each individual must fulfill in society. It emphasizes responsibility towards the environment, fellow beings, and future generations. In the ancient context, kings and rulers were expected to uphold Dharma by ensuring justice, resource-sharing, and public welfare.	Corporate Social Responsibility (CSR): The concept of Dharma aligns with modern corporate sustainability frameworks. Businesses can integrate ethical values into their operations, ensuring that their production processes are environmentally friendly, employees are treated fairly, and local communities' benefit. Environmental Governance: Governments can use Dharma as a guiding principle to enact policies that balance economic growth with environmental justice. This can lead to eco-friendly policies, such as carbon trading, environmental protection laws, and equitable resource distribution.
The Cycle of Life and Circular Economy Ancient Indian philosophies, especially in Hinduism and Buddhism, emphasize the cycle of life (birth, death, and rebirth) and interdependence among all elements of nature. This cyclical view of life stresses that nothing is wasted, and everything returns to the cycle for rejuvenation.	Circular Economy: The idea of the cycle of life translates well into the circular economy model, which promotes resource reuse, recycling, and minimizing waste. Traditional practices like composting and upcycling can be integrated into modern waste management systems to promote zero waste and material efficiency. Sustainable Manufacturing: Industries can adopt the reduce-reuse-recycle model, using raw materials that are sustainably sourced and designing products that are fully recyclable at the end of their life cycle.

Community -Based Conservation (Panchayat system) The Panchayat system, a community-led governance model, played a central role in the management of common resources like forests, water bodies, and grazing lands. These community-based councils enforced rules based on collective wisdom and local needs.	Community-Led Conservation: Today, Panchayats can be revived for local governance over natural resources and environmental decision-making. Communities can take part in forest protection, water management, and sustainable land use policies. Participatory Planning: This ancient practice encourages bottom-up conservation efforts, where local communities are actively involved in environmental planning. Modern environmental programs can adopt decentralized decision-making, empowering communities to lead local conservation initiatives.
Ahara (Diet) and Sustainable Living Ahara (diet) in ancient texts like the Ayurveda emphasized consuming food that was locally sourced, seasonal, and in harmony with one's body and environment. Foods like grains, pulses, vegetables, and fruits were considered sustainable and nutritious.	Plant-Based Diets and Sustainable Food Systems: The principle of Ahara aligns with modern trends promoting plant-based diets and the reduction of meat consumption, which significantly lowers carbon emissions and water usage in food production. Local, Seasonal Eating: Promoting local and seasonal foods reduces the carbon footprint of transportation, packaging, and waste associated with global supply chains. Modern food systems can use traditional knowledge to promote sustainable food production and eco-friendly agriculture.

Water Conservation and Sacred Water Bodies Ancient Indian practices viewed water bodies (rivers, lakes, ponds) as sacred and **connected them to the divine**. Rituals, festivals, **and** temple activities **often focused on** water preservation **and** cleanliness.	**River Rejuvenation**: Sacred rivers like the **Ganga** are now part of government schemes such as **Namami Gange**, aimed at restoring the **sanctity and cleanliness** of these water bodies through **pollution control** and **reforestation**. **Community Involvement in Water Conservation**: Engaging local communities in water protection, using **ancient practices** like **rainwater harvesting** and **water body protection**, is crucial to meeting **modern water scarcity challenges**.

Source: Secondary Source

Conclusion

Deep insights into living in balance with environment may be gained from ancient Indian ideas, and these insights can be successfully applied to contemporary sustainability issues. We can handle the climate catastrophe and manage the environment in a more sustainable and balanced way by fusing scientific advancements with spiritual understanding. By combining old wisdom with contemporary conservation impacts, India can build a sustainable future anchored in its cultural past. Traditional Indian knowledge was naturally sustainable, guaranteeing that social and economic advancement did not come at the expense of environmental deterioration. India—and the world—can build a really sustainable future where growth is in line with ethics, well-being, and the environment by bridging the gap between the past and present.

Findings

- Environmental preservation was ingrained in everyday life, ethics, and religion.
- The Cycle of Life, Dharma (duty), and Ahimsa (non-violence) were among the principles that highlighted one's obligation to the natural world.
- By means of spiritual and cultural traditions, sacred trees, water sources, and hotspots for biodiversity were preserved.
- The Panchayat system made sure that natural resources were managed locally.
- Water security was guaranteed by community-led water conservation measures including rainwater collection, stepwells, and temple tanks.
- Village-level management and regional customs like Shrenis (guilds) aided in the sustainable management of water, agriculture, and forests.
- Soil health and biodiversity were guaranteed by conventional farming practices such crop rotation, organic agriculture, and mixed cropping.
- Utilising organic insecticides (Neem, Turmeric, Panchagavya) and natural fertilisers (cow dung, green manure) decreased environmental deterioration.
- Traditional agricultural methods protected soil fertility and avoided water contamination, in contrast to contemporary chemical-based agriculture.
- Sacred rivers and temple tanks were socially safeguarded, limiting water abuse and pollution; ancient water management methods like Karez (underground canals), Johads (rainwater gathering ponds), and Baolis (stepwells) efficiently saved water.
- These techniques may be used to modern river revitalisation initiatives (like Namami Gange) to

increase sustainability.

- The zero-waste concept of ancient Indian customs involved recycling and reusing materials.
- It was common practice to compost, upcycle used materials, and use sustainable packaging (banana leaves, clay pots).
- Reviving historic waste management and circular economy concepts can help alleviate the current trash challenge.
- Yoga and Ayurveda promoted eco-friendly lifestyles, seasonal diets, and natural cures as key components of sustainable living.
- By encouraging plant-based, locally grown, and seasonal foods, Sustainable Diets (Ahara) helped to lower carbon footprints and increase food security.
- This is in line with contemporary sustainable food trends such as slow food consumption, organic food, and farm-to-table.

- **Suggestions**
- Traditional environmental knowledge should be included into government policies and sustainable development plans.
- Local communities may be empowered to preserve forests, water, and land by reviving panchayat-led natural resource management.
- Promote local involvement in conservation by using decentralised governance forms, such as the Panchayats of the past.
- Reintroduce conventional methods of conserving water in contemporary infrastructure projects (temple tanks, rainwater gathering).
- Encourage neighbourhood-based afforestation initiatives and incorporate them into national forest

strategies.

- Using conventional methods like crop rotation, natural fertilisers, and insect control, farmers are being trained in sustainable agriculture.
- Promote the adoption of conventional water-efficient irrigation techniques to lessen reliance on groundwater depletion.
- Vastu-inspired sustainable designs, with a focus on water-efficient layouts, green structures, and solar passive cooling, should be included into urban planning
- To raise understanding of traditional conservation methods, include old Indian environmental philosophy into school curricula.
- For greater accessibility, digitise classic environmental works such as Arthashastra, Vrikshayurveda, and Vastu Shastra.

REFERENCES

- **Books & Academic Sources**
- **Kautilya's Arthashastra** – R. Shamasastry
- **Environmental Ethics in Ancient India** – O.P. Dwivedi
- **Sustainability in Ancient India** – Radhakrishna Choudhary
- **Jain Ecology: The Environmental Ethics of Jainism** – Michael Tobias
- **Ancient Indian Agriculture and Environment** – Nandita Krishna
- **Research Papers & Journals**
- Bhardwaj, R. (2018). *Ancient Indian Environmental Ethics and Their Relevance Today*. Journal of Ecological Studies, 15(2), 45-60.

- Chatterjee, P. (2021). *Religious Festivals and Environmental Awareness in India: A Comparative Analysis*. Indian Journal of Sustainable Studies, 8(1), 22-37.
- Desai, A. (2021). *Traditional Water Conservation Systems in India and Their Modern Relevance*. Environmental Policy Journal, 12(3), 118-132.
- Gupta, S. (2020). *Indigenous Knowledge and the SDGs: A Case for Integrating Traditional Environmental Practices into Policy Frameworks*. Global Sustainability Review, 9(4), 75-91.
- Kumar, V. (2017). *Ahimsa and Environmental Conservation: Insights from Jain and Buddhist Traditions*. Journal of Environmental Philosophy, 10(2), 98-112.
- Mehta, R. (2017). *Sacred Groves in India: A Model for Community-Driven Conservation*. Indian Journal of Biodiversity, 7(3), 55-70.
- Mukherjee, P. (2020). *The Atharva Veda and Ecological Consciousness: A Historical Perspective*. Vedic Studies Review, 6(1), 39-55.
- Prasad, K. (2016). *Sustainable Agriculture in Ancient India: Lessons from the Past*. Agricultural History Journal, 14(2), 66-81.
- Rajagopal, L. (2019). *Community-Led Conservation in Sacred Groves of Maharashtra and Karnataka*. Ecology and Society Journal, 11(2), 87-102.
- Rao, M. (2021). *Mainstreaming Indigenous Knowledge in Modern Environmental Policies: Challenges and Opportunities*. Policy Studies Journal, 13(4), 123-138.
- Reddy, B. (2020). *Environmental Protection in Hindu Traditions: The Role of Rituals and Beliefs*. Cultural Ecology Journal, 5(1), 29-45.
- Sharma, P. (2015). *Dharma and Ecology: A Historical*

Study of Indian Environmental Ethics. Asian Philosophical Journal, 9(3), 103-118.

- Shankar, G. (2023). *National Green Tribunal and Traditional Environmental Knowledge: A Review of Policy Integration.* Indian Policy Review, 10(1), 56-72.
- Singh, N. (2022). *Chakra Vyuh and Circular Economy: Ancient Indian Economic Planning for Modern Sustainability.* Economic Development Journal, 8(2), 47-65.
- Tripathi, A. (2018). *Reviving Stepwells and Johads: Sustainable Water Management in Rajasthan.* Water Conservation Research, 7(3), 89-104.
- Verma, T. (2022). *Integrating Traditional Knowledge into National Sustainability Policies: A Case Study of India.* Journal of Environmental Policy, 9(3), 112-128.
- **Government Reports & Policies**
- **National Mission for Sustainable Agriculture (NMSA)** – Ministry of Agriculture, Government of India.
- **Jal Shakti Abhiyan** – Ministry of Water Resources, India.
- **India's ESG (Environmental, Social, Governance) Guidelines** – SEBI Report.
- **Sacred Groves & Biodiversity Conservation** – Report by Wildlife Institute of India.
- **Namami Gange: A Case Study on River Conservation** – Ministry of Environment, Forest, and Climate Change.

Integration of community-based cultural control practices in compacting Pests in the Rice ecosystem

Dr Gayathri Elayidam U

*Associate Professor in Zoology, VTM NSS College,
Dhanuvachapuram, Kerala India*

Abstract

Insects and pests pose a significant hazard to global food production, with India being the largest rice crop. Diseases, insects, and weeds cause 120-200 million tons of rice loss yearly. The ecosystem of rice fields is dynamic, and traditional knowledge systems are used to combat pests. However, community-wide measures are only effective if carried out over 50 areas. Traditional pest management techniques are environmentally safe, but indigenous knowledge has progressively disappeared. Therefore, it is vital to gather and record this information for efficient farming practices in integrated pest management plans.

Key Words:*Oryza sativa*, IKS, Cultural control practices, IPM.

Introduction

Numerous insects and pests target crops, which harms global food production. According to the WHO, pesticides are chemical substances used to eradicate undesired plants, insects, rodents, and fungi. The Environmental Protection Agency (EPA) defines pesticides as insecticides, herbicides, fungicides, and other chemicals used to manage pests that are harmful to humans and other species and must be handled and disposed of appropriately. According to Kumar et al. (2012), India is Asia's second-largest producer of pesticides. India has been listed as one of the top pesticide-using nations in several reports (Gupta, 2004). The crop that uses pesticides the most worldwide is rice. To highlight some alternatives to these lethal chemical pesticides and the significance of using Integrated Pest Management systems and Good Agricultural Practices, this review attempts to analyze cultural control practices

The ecosystem surrounding rice fields is a highly fluctuating manmade wetland, and as the agricultural cycle shifts, it also changes its natural habitats. The environment of rice fields is composed of up of marshy areas, dry land, soil, air, and aquatic habitats. These regions form an ecological niche with a wealth of biological diversity. A survey performed in the ricefields of Bathalagoda, Kurunegala District, indicated a faunal richness of 103 vertebrates from five classes and 495 invertebrates from ten taxa. There are approximately 82 species of macrophyte plants, encompassing Pterydophytes, broad-leaved weeds, and grasses (Jayanthi and Channa 2006). A range of living forms can find homes in the small-scale environments created by the rice's various stages.

Chloroperiphos, Danitol, Cartap, Padaan, Imidacloprid, Alfa, Beta, Gamma Hch, Aldrin, Dieldrin, Heptachlor,

Malathion, Malaxon, Phosalone, Phosphomidon, Pyre-thrins, LambdaCyhalothrin, Hexaconazole, Deltamethrin, Permethrin, Anilophos, Bifenithrin, Carbariyl, Carbofuron, Sulfosulfuron, 2,4-D, and Mcpa are a few of the pesticides that are frequently used in Indian rice fields. DDT, endrin, endosulfan, folidon, chlorpyrifos, monocrotophos, quinal-phos, karate furadan, and other pesticides are used on rice in our state of Kerala, according to a 2011 research by Thanal, an NGO in Kerala. They are poisonous to all living things, including humans, and extremely persistent. Orissa, Kar-nataka, Kerala, and Tamil Nadu are among the states that employ the following important pesticides in paddy: car-baryl, carbofuran, chlorpyrifos, endosulfan, lambdacyhalo-thrin, malathion, monocotrophos, and phorate. According to the herbicide Action Network (PAN), every herbicide on the list is considered a Highly Hazardous Pesticide (HHP).

The insecticides are available as dusts, wettable powders, emulsifiable concentrates, and liquids. When pesticides are applied, they travel through the air and wind up in different areas of the ecosystem. Pesticides seep into the soil, groundwater, and surface water. Pesticides stay as residue in non-target plants and animals after entering the food chain (Rajendran.S, 2003). According to Yadav et al. (2015), certain pesticides, such as aldrin, chlordane, DDT, dieldrin, endrin, heptachlor, and hexachlorobenzene, contain persistent organic pollutants (POPs) that do not break down and can linger in the environment for years. Chemicals can bioaccumulate, biomagnified, and bioconcentrate up to 70,000 times their initial concentration (Hernández et al., 2013a).

Creation of Rice Pest Management Strategies

There has been a change from a largely unilateral approach to insect control, which heavily relied on insecticides, to a multilateral approach that involves a variety of control strategies as rice scientists and farmers have gained experience in growing the modern varieties and the agronomic practices that have accompanied the "Green Revolution." "A broad ecological attack combining several tactics including biological, chemical, and cultural control methods and insect resistant rice varieties, for the economic control and management of pest populations" is the simplest way to describe this strategy, which is known as integrated pest management (IPM).

Cultural Practices

Altering production methods to create an environment that is less conducive to pest invasion, reproduction, survival, and dissemination is known as cultural control. Its objective is to reduce pests. The primary goal of cultural control is to make the environment less conducive to pests and more conducive to their natural enemies. This can be achieved by either directly affecting the growth and multiplication of insect pests or by reducing the likelihood that they will attack plants. are carried out especially for crop husbandry, including weeding and land preparation, but they also serve to reduce insect accumulation.

Crop production techniques that serve the dual purpose of crop production and pest suppression are cultural approaches to insect management. These methods have been passed down through the generations and were created by farmers via many years of trial and error. For example, primary cultural management approaches include draining a field to reduce aquatic caseworm larvae or planting a trap crop to suppress stem borers (Litsinger, J.A. 1994)

Fundamentals of Rice Insect Pest Cultural Control

Crop yield and pest control are the two main goals of the majority of traditional agronomic techniques. Transplanting seedlings to establish rice is a good example of how to successfully suppress weeds and lower the prevalence of seedling pests. These methods have been created by farmers via trial and error and observation. These methods of producing crops are passed down from one generation to the next and typically use farm-based technologies with minimal reliance on outside resources (Litsinger, J.A. 1994). Through intricate interactions with the crop and environment, yield is ultimately impacted by changes made to any agricultural production method. Insect pest populations are positively or negatively impacted by all crop production methods. (Reissig *et al.*, 1986)

Different insect species may respond differently to a particular technique, such as planting time or plant spacing. They often balance each other out. Although draining the fields or applying low fertilizer rates are quite successful at controlling some insect pests, they may also result in a decreased yield. Developing cultural control techniques necessitates a deep understanding of the life cycles and environments of both the insect and its plant host. Farmers must determine which cultural practices— like direct seeding or transplanting seedlings in a wetland environment—are appropriate for their area.

There is a wealth of folklore about indigenous cultural activities in rice culture. Among the cultural control practices that may help control rice insects are: (1) mixed cropping; (2) planting methods (direct seeding vs. transplanting); (3) age of seedlings at transplanting; (4) water management; (5) fertilizer management; (6) crop rotation; (7) number of rice crops annually; (8) planting time; (9) synchronous vs.

asynchronous planting over a given area; (10) trap crop; (11) tillage; (12) weeding; and (13) crop growth duration.

How to Use Cultural Practices to Control Weeds?

1. Timing: From the time of planting until the crop canopy closes, weeds must be managed.
2. Leveling and land preparation: Use leveling and land preparation to suppress expanding weeds and promote the germination of weed seeds. Repeat tillage at sufficient intervals (~10 days) to eradicate newly sprouting weeds.
3. Cut down on weeds entering fields: Avoid introducing weeds into fields by: a) using high-quality, clean seed; b) keeping weeds out of seedling nurseries so that rice seedlings are not planted alongside weeds; c) keeping irrigation channels and field bunds free of weeds to keep weed seeds or vegetative parts out of the fields; d) using clean equipment to avoid contamination of the fields or crops; and e) rotating crops to break weed cycles. f) Fallow management: To stop flowering, seed-set, and the accumulation of weed seeds in the soil, eradicate weeds in fallow fields (for example, by tillage)
4. Weed-crop competition: To suppress weeds, choose a weed-competitive cultivar with high tillering and early seedling vigor. Compared to straight-seeded crops, transplanted crops often have fewer weeds and less yield loss. In the early stages, transplant robust, healthy seedlings that can more effectively compete with weeds. Keep a sufficient number of plants that will shade out weeds by closing their canopy by tillering as much as possible. To reduce rice-weed competition for nitrogen (N), apply nitrogen (N) fertilizer immediately after weeding.

5. Water management: The most effective weed control is water. Most grasses and certain sedges are among the many weeds that cannot sprout or flourish in wet circumstances. To reduce weed pressure and prevent weed emergence, keep the field's water level between 2 and 5 cm. Fields can be constantly flooded from the moment of transplanting until the crop canopy completely covers the land if there is enough water. To prevent high areas where weeds can establish themselves, the field must be leveledproperly.(http://www.knowledgebank.irri.org/training/fact-sheets/pestmanagement/weeds/cultural-weed-control)

Examples of Indigenous Knowledge system in Rice Cultivation

- **Folk Songs & Proverbs:** Farmers use songs (e.g., KarisalPadal in Tamil Nadu) to share knowledge about rice farming.
- **Apatani Tribe (Arunachal Pradesh):** The Apatani tribe in Arunachal Pradesh has developed a remarkable system of soil and water management for wetland rice cultivation, emphasizing the conservation of surrounding forests and the recycling of agricultural wastes (https://www.fao.org/climate-smart-agriculture-sourcebook/production-resources/module-b7-soil/b7-case-studies/case-study-b7-3/fr/)
- In Kerala, traditional rice varieties and cultivation practices are being actively conserved and promoted by organizations like the Centre for Indian Knowledge Systems (CIKS), which also focuses on organic farming and sustainable agriculture (https://ciks.org/)
- The thirty indigenous practices identified were documented by classifying them into rational and

irrational practices based on the evaluation of a group of scientists and agricultural extension officials working in the area (Binnoo and Vijayaraghavan 2001)

Modern Applications and Reviving Traditional Wisdom

- Integration of IKS into **agroecology and permaculture**.
- Use of **GIS and AI** for mapping indigenous rice biodiversity.
- Revitalization of **forgotten rice varieties** through community seed banks.

Conclusion

The Indian Knowledge System (IKS) and Cultural Control System in rice cultivation reflect a blend of traditional wisdom, ecological understanding, and socio-cultural practices that have been passed down through generations. These systems ensure sustainable farming, biodiversity conservation, and food security. The Indian Knowledge System in rice cultivation showcases an eco-friendly, sustainable, and holistic approach. Blending this wisdom with modern scientific research can ensure food security and environmental conservation.

Reference

- Binoo P. Bonny &K. Vijayaragavan (2001) Evaluation of Indigenous Knowledge Systems of Traditional Rice Farmers in IndiaJournal of Sustainable Agriculture 18(4):39-51 DOI:10.1300/J064v18n04_06
- Gupta, P.K. (2004) Soil, Plant, Water and Fertilizer Analysis. Agro Botanica
- Rice fields: an ecosystem rich in biodiversity
- Hernandez, A. J.; Roman, D.; Hooft, J.; Cofre, C.; Cepeda, V.; Vidal, R., 2013. Growth performance and

expression of immune-regulatory genes in rainbow trout (*Oncorhynchus mykiss*) juveniles fed extruded diets with varying levels of lupin (*Lupinus albus*), peas (*Pisum sativum*) and rapeseed (*Brassica napus*). Aquacult. Nutr., 19 (3): 321-332

- Kumar, D. S.; Prasad, R. M. V.; Kishore, K. R.; Rao, E. R., (2012). Effect of Azolla (*Azolla pinnata*) based concentrate mixture on nutrient utilization in buffalo bulls. Indian J. Anim. Res., 46 (3): 268-271
- Litsinger, J.A. 1994. Cultural, mechanical, and physical control of rice insects. In book: Biology and Management of Rice Insects. (pp.549-584) 6Wiley Eastern Ltd., New Delhi, 779 p. Editors: E.A. Heinrichs
- Jayanthi P. Edirisinghe''' and Channa N.B.(2006) Bambaradeniyj. Natn.Sci.Foundation Sri Lanka 34(2): 57-59
- Rajendran, S. (2003) Grain Storage: Perspectives and Problems. In: Handbook of Postharvest Technology: Cereals, Fruits, Vegetables, Tea and Spices, Marcel Dekker, New York, 183-192.
- Reissig WH, Heinrichs EA, Litsinger JA, Moody K, Fiedler L, Mew TW and Barrion AT., 1986. Illustrated Guide to Integrated Pest Management in Rice in Tropical Asia. International Rice Research Institute, Philippines, 411p
- Yadav et al. (2015), Cultural, mechanical and physical control of rice insects. Pp. 549-584 In: EA Heinrichs (ed.) Biology and Management of Rice Insects. International Rice Research Institute, Philippines, 779p.

Web resources
- http://www.knowledgebank.irri.org/training/fact-

sheets/pest-management/weeds/cultural-weed-control
- https://www.fao.org/climate-smart-agriculture-sourcebook/production-resources/module-b7-soil/b7-case-studies/case-study-b7-3/fr/
- https://ciks.org/

Application of Mathematics in Astronomy: A Descriptive Study

Dr. Remya Stanley

Assistant Professor,
Mount Tabor Training College, Pathanapuram

Abstract

Math and astronomy have a long and interconnected relationship, and they are still closely linked today. New planets or celestial bodies are found every year, possibly requiring the application of mathematics. Astronomers design mathematical models that describe the formation, history, and future of celestial bodies that are as accurate as possible. This study explores the application of Mathematics in Astronomy.

Keywords: mathematics, astronomy

Introduction

Astronomy, as it is defined, is a natural science involving the study of celestial objects and phenomena, as well as the known existing universe. More specifically, of the Milky Way Galaxy, the galaxy we currently live in and as such, have the most knowledge of, as well as other officially named galaxies that scientists are still studying, i.e. the Andromeda Galaxy. On the other hand, mathematics is the study of various properties of quantities, shapes, and formulas. Both subjects have varying sub-categories such as cosmology, astrophysics, geometry, calculus, etc. Math is present in many aspects of everyday life, and science is no exception. Astronomy is also an example of such a science, and it shares an interesting relationship with math that is worth being explored. Exploring this relationship between the two will grant a better understanding of them in the world both in the past and today.

Need and significance of the study

Mathematics was also an important component of astronomical sciences, as most of the ancient Indian scholars dealt with numbers and equations in order to make astronomical assumptions. These texts include, among other things, one of the first expressions of the idea behind what is now known as Pythagoras' theorem in geometry (the ancient Babylonians were also aware of this principle). The earliest evidence of ancient Indian astrological knowledge may be found in Vedanga books on Jyotish or astrology, the major purpose of which was to determine the dates of sacrificial ceremonies. It, later on, evolved from astrology to astronomy or "Khagol Shastra" when scientific studies were gaining momentum in Ancient India. The Yavanajataka, a Sanskrit treatise, documents the introduction of Hellenistic concepts of astronomy and mathematics in India. There are a few number of studies were conducted to find out the application of Mathematics in Astronomy. The investigator is interested to find out the application of Mathematics in astronomy. Hence the need and significance of the study.

Objective - To find the application of Mathematics in Astronomy

Method - Qualitative method

Tools - Secondary resources

Findings

Astronomers have always used mathematics to study the stars and planets. The ones of old have used math, and the same is true for astronomers today. Astronomers use math to calculate the distance of the travel of space shuttles and the like, to ensure it arrives at the space station with no damage or mishaps. Ancient astronomers also used mathematics for the calculation of distance regarding navigation or travel. Most of the math used is concerning spacecraft and relative distances of celestial objects. The findings of the study are the following.

Applications of Mathematics in Astronomy

Primitive societies may have used arithmetic to keep track of lunar and solar cycles and animals, food, and people. However, the first mathematical, and astronomical innovation was made during the Mesopotamian and Babylonian civilizations. The sun's apparent motion was used to predict eclipses and celestial body positions in terms of the degree of latitude and longitude.

The story of mathematics grows fascinating as we approach one of humanity's most successful geniuses. Sir Isaac Newton devised Calculus as he studied Halley's Comet. This way of addressing moving bodies allowed him to replicate the movement of not just Halley's Comet perfectly but any other celestial body that traveled across the sky.

Astronomers frequently use math in their work to attain their objectives. They use it to determine the paths of satellites, rockets, and space vehicles, as well as to convey signals in the global positioning system when compressed data is sent. Algebra is used to compute speed and track motion. Without it locating devices like the Hubble Space Telescope will be difficult.

In space research missions, astronomers use technology and math, which has led to substantial discoveries. Earlier astronomers used to hire mathematicians to help them with complicated calculations. Consider this scenario: an astronomer using a telescope to see celestial objects captures a sequence of numbers on the telescope's camera. These figures refer to the total amount of light emitted by various celestial objects - stars, star clusters, etc. Now, in order to comprehend these figures, arithmetic and statistics are required.

One of the inventors of mathematics, Pythagoras of Samos, concluded that each planet is connected with spheres. Ptolemy, who recorded longitudes and latitudes, developed an earth-centered mathematical model of the Solar system. Kepler studied the orbits of the planets mathematically and later discovered the laws of planetary motion. In addition, Isaac Newton explained gravity and described how planets move about one another, and today, his equations are used to calculate gravitational forces. While studying objects in motion, Galileo observed that the speed with which a heavy object falls is not directly proportional to its weight.

Astronomy is a mathematical discipline. Today, astronauts use arithmetic to navigate a space shuttle back to Earth for landing. To avoid collisions between fast-moving objects at a single point, which might result in harm to one another, complex mathematical calculations must be done.

Astronomers, unlike geologists, cannot go into the field or perform tabletop experiments as physicists.

So they rely on logic, reason, and facts. Since their only source of data is a galaxy 40 million light-years distant, astronomers must be mathematicians and extract every bit of information from each observation. Astronomers use math in various ways, including the creation and development of laws that regulate celestial objects. When we look at things in the sky via a telescope, the telescope's lens captures a series of numbers that show how much light and what kind of individual light objects emit, and so on. Arithmetic and statistics are used to understand the data collected.

Most mathematical, astronomical operations, such as spherical trigonometry, which is based on data received from an observer on Earth, are focused on the location and calculation of relative distances of celestial bodies. The capacity to project the celestial sphere onto a flat surface set the stage for developing instruments like the astrolabe and sky charting. For instance, in northern France, a twelfth-century monk aligned stars with historical monuments in his monastery, such as the windows along the dormitory wall. Astronomy grew more accurate as precise mathematical algorithms were created.Many of the world's greatest astronomers were also mathematicians and vice versa. Hence, astronomical calculations have influenced and inspired mathematical breakthroughs. As soon as the data is quantified, astronomers can compute and predict observations.

Celestial Linkage

Mathematics is much more than a collection of fuzzy equations and complicated rules. Mathematics is a universal language and understanding it allows you to access the fundamental systems that govern the universe. It's identical to visiting a new country and gradually learning the native language so that you may learn from them.

As species restricted to our solar system, this mathematical endeavor is what permits us to dive into the depths of the cosmos. Currently, there is no method for humans to go to the core of our galaxy and confirm the presence of the black hole. We cannot witness a star evolve in real-time in a Dark Nebula. However, thanks to mathematics, we can grasp how these objects exist and operate. When you start learning arithmetic, you are not only growing your intellect, but you are also making a basic connection with the cosmos. You may investigate physics at the event horizon right from your computer.

Conclusion

Math is the engine of the universe, driving and dictating all actions, interactions, transformations. It forms the basis of science, a tool for exploring and explaining the universe in a language that is common to everyone and not just a few. Mathematics ranks as the most pre-eminent tool in astronomy. The epic tale of the cosmos is written in numbers, and our capacity to interpret those numbers into events that we all enjoy learning about is nothing short of phenomenal. So, if you get the chance to learn math, take advantage of it since math connects us to the stars.

Bibliography

- Brown,I,F. (2022). Astronomy and Mathematics. Retrieved from https://soar.suny.edu/bitstream/handle/20.500.12648/12151/4516_Imani_Fouchong-Brown.pdf?sequence=1&isAllowed=y
- http://www.sites.hps.cam.ac.uk/starry/mathematics.html

- http://www.sites.hps.cam.ac.uk/starry/mathematics.html
- https://www.byjusfutureschool.com/blog/how-math-and-astronomy-are-related/
- Sharma et.al. (2024). Aryabhata's Enduring Contributions to Astronomy in Ancient India: Unveiling the Cosmic Secrets. IJFMR240324020. International Journal for Multidisciplinary Research (IJFMR). 6(3). E-ISSN: 2582-2160

Investigating the Possibilities for Sustainable Food Systems and Agricultural Sustainability in Kerala through the Cultivation of Millets in the Chinnar Wild Life Sanctuary

Mithraja M J[1], Kavitha K R[2]
Sushama Raj R V[3*]
[1]Mahatma Gandhi College, Thiruvananthapuram University of Kerala, [2]S.N. College, Chempazhanthy, Thiruvananthapuram, University of Kerala. [3]VTM NSSCollege, Dhanuvachapuram, Thiruvananthapuram, University of Kerala.
[*]Corresponding Author: drsushamarajrv@gmail.com

ABSTRACT

The Chinnar Wild Life Sanctuary is located in the Idukki district of Kerala, India, in the Western Ghat mountain range (Latitude 10.3ºN, Longitude 77.2ºE). Millet, a staple crop in Kerala, has several benefits for food security, climate resistance, and soil health. It is a crucial component of agriculture that is sustainable. The research project assesses how it could promote sustainable farming in Chinnar, Kerala, a region renowned for its harsh climate fluctuations and fragile ecosystems. This study aims to evaluate the potential benefits of millet farming in reducing the effects of climate change, improving soil health, enhancing biodiversity protection, and increasing food security while using less water. The study employed a combination of methods, such as field research, surveys, and focus groups. the capacity of millet cultivation to preserve biodiversity and offer resilience to climate change. The results show that millet production provides a consistent supply of nutrients, decreases soil erosion, and enhances soil health by adding organic matter. By providing a home for insects and microbes, it also aids in biodiversity conservation. We gathered and analyzed a variety of millets for this study, including the Thina varieties of pulluthina, mulianthina, kambanthina, the Chama varieties of Periya chama, karusama, and vella chama, and the Varagu types of vellavaragu, Karuvaragu, and kuthira vali (vernacular names). According to the study's findings, promoting millet production in Chinnar, Kerala, can significantly improve agricultural sustainability. Millets can preserve biodiversity, promote food accessibility, improve soil health, and consume fewer resources, according to studies. According to the study's conclusions, further research is necessary to look into the problems of soil erosion, biodiversity loss, and climate change.

Key Words: - Millets, sustainable development, Chinnar, Chama, Thina, Varagu

Introduction

Millets are nutrient-rich grains belonging to various subfamilies of the Poaceaefamily. The UN General Assembly adopted a resolution declaring 2023 as the International Year of Millets (Kheya et al, 2023). Millets are tiny grains typically cultivated in arid and tropical areas of Africa and Eurasia. They belong to the Poaceae family. Evidence indicates that millet was consumed as early as 3000 BC during the Indus Valley Civilization, rendering it one of the earliest domesticated crops. They are ranked fifth among the most important cereal grain crops globally, following wheat, rice, maize, and barley (Antony et al, 2022). Millets are designated as "grains of the impoverished" or farm cereals (Babele et al, 2022). These grains are cultivated due to their necessity for human consumption and livestock feed. Millets bolster rural economies, promote sustainable resource utilization, preserve biodiversity, and enhance food security. (Choudhary et al., 2023). Millets serve as a staple meal, particularly in the semiarid and economically disadvantaged tropical regions of Asia and Africa (Prabhu and Gowri, 2017). Michael and Shanmugam (2013) assert that millets comprise a varied array of smallseeded grasses cultivated globally as cereal crops or grains for human consumption and animal feed. Millet is a significant factor in the advancement of more resilient and sustainable agricultural practices (Mrabet, 2023).

In this context, millet exemplifies the diversity, adaptation, and resilience of agriculture. This topic is intriguing for exploration in sustainable agriculture due to its historical importance and distinctive characteristics (Magdoff et al 2010). Future advancements may generate climate-resilient, high-yielding millets that are more suitably adapted to diverse agroecological conditions. To enhance millet productivity, features such as shortened growth cycles, elevated nutritional value, and insect resistance may be integrated (Srivastava et al., 2023). These technologies assist farmers in making educated decisions and optimizing agricultural operations for enhanced productivity and sustainability by delivering real-time data on crop health, soil conditions, and resource consumption (Tripathi et al., 2023). Millets are highly nutrient-dense and a superior source of proteins, vitamins, and minerals. Animal feed and alcoholic beverages are produced from the remaining twenty percent of millet grains. Approximately 80 percent of millet grains are utilized for consumption. Padulosi et al. (2015); Rao (2007).

To assess the environmental sustainability of millet farming including its water use, biodiversity, and soil health. To evaluate the economic significance of millets by examining market demand and price volatility. To examine the communal impact of millet planting. Foster the sustainable cultivation of millets among indigenous tribes and others through daily routines.

MATERIALS AND METHODS
Description of the Study Area

Chinnar Wildlife Sanctuary extends from latitudes 10º15′ to 10º21′ N and longitudes 77º5′ to 77º16′ E. Gamble, J. S., & Fischer, C. E. C. (1956) The 16-kilometer SH 17 Munnar – Udumalpet route divides the sanctuary into nearly equal portions. The region, situated in the rain shadow of the southern Western Ghats, receives an average annual rainfall of approximately 500 mm, distributed over roughly 48 days. Eco-Informatics Centre, Chinnar Wildlife Sanctuary. The Chinnar Wildlife Sanctuary consists of eleven tribal communities inhabited by around 1,800 individuals. The predominant members of the Muthuvan and Hill Pulayas communities residing in the villages of Thayannankudy, Mulangamutty, Vellakkal, Puthukkudy, Eiruttalakkudy, Eechampetty, Alampetty, Palappetty, Champakkad, Mangappara, and Ollavayal are engaged in small-scale agriculture.

Field Study

1. The survey used a qualitative method to obtain information from selected farmers belonging to the Kanis or tribal communities, utilizing a questionnaire to directly gather their traditional knowledge.
2. Group talks were led by the chieftain of each tribal community, focusing on traditional knowledge and information regarding the social and economic strategies of millet growing.
3. Field observations were conducted to gather information from different tribal communities inhabiting Chinnar Wildlife Sanctuary villages to collect different samples.

Result and Discussion

The environmental study indicates that millet farming enhances soil health and diminishes soil erosion. Millet farming necessitates a less quantity of water in comparison to other crops. This facilitates water conservation and promotes biodiversity, so benefiting birds, insects, and microorganisms.

The tribal inhabitants of Chinnar harvest millet on their own land employingtraditional techniques, without the aid of machinery. The harvesting and dissemination of seeds fulfill cultural ceremonial roles. Farmers mostly utilize indigenous varieties of millets that they have gathered over the years. All millets utilized in the community have been preserved by the chieftain (Oorumooppan). Various varieties of thina (foxtail millet) have been assigned regional designations, including thina, Pulluthina, Mulianthina, and Kamban thina. Four kinds of tiny millet (Chama) are Periyasama, Karu sama, Vella sama, and Chama, along with Varagu (Paspalum) variants such as karuvaragu, vellavaragu, and Kuthira vali (Echinochloa). Nearly all tribes utilize millets as sustenance, fodder for their livestock, and some employ them for medicinal purposes. The diverse array of millets cultivated by traditional techniques and indigenous knowledge renders the Thayyannakudy tribal hamlets particularly noteworthy. Each native variety has been conserved for future production.

The Punarjeevanam study (2016) extensively examined millet types in Chinnar Wildlife Sanctuary. These millets will be retained as necessary due to their prominence and iconic status in this village. The indigenous population promotes the utilization of this millet due to its benefits for women's and children's health. The use of indigenous millets improves the sustainability of traditional millets, cultural heritage, and quality of life in the hamlets. Millet is considered a superfood due to its gluten-free nature, allowing anyone with gluten sensitivities or allergies to consume it safely. Millet grain flours are utilized to produce bread, biscuits, chapatis, and porridge to enhance micronutrient deficits (Ramashia et al., 2021).

Future Perspectives

Millets, offering resilience to climate change, nutritional benefits, and opportunities for rural economic development, possess the capacity to significantly enhance future food security, empower women, and bolster rural livelihoods, enabling farmers to produce high-quality millet for the market and increase their income. Future investigations into the stress tolerance of millets may employ fresh genetic and molecular methodologies to identify significant genes and regulatory networks associated with stress.

Conclusion

The examination of hamlets within the Chinnar Wildlife Sanctuary in Kerala constitutes the most significant aspect of our research. The variety of traditional millets within the tribe Paniceae is significant to the cultural system. The native foods, fodder, and medical knowledge related to millets have been documented with the assistance of the elderly tribal members in the hamlets. They were cognizant of all environmental circumstances, including field and weather factors, which aids in mitigating losses in crop cultivation. It aids in the meticulous preservation of traditional millets for the future utilization of indigenous communities. The extensive genetic diversity of millets in the research area is a significant asset for agricultural advancement and ecological conditions.

Acknowledgement

The authors express gratitude to the Head of the Department of Botany at Mahatma Gandhi College, Kesavadasapuram, Kerala, for facilitating the research study. The authors express gratitude to the Head of the Department of Forest, Kerala, and the Directorate of Scheduled Tribe Development for granting permission to access the forest and tribal hamlets. The authors express gratitude to the University of Kerala for the financial support provided.

Conflict of Interest: None

Authors Contributions: All authors of this paper have directly participated in the planning, execution, analysis of this study.

References

- Antony Ceasar, S., & Maharajan, T. (2022). The role of millets in attaining United Nation's sustainable developmental goals. *Plants, People, Planet, 4*(4), 345-349.
- Babele, P. K., Kudapa, H., Singh, Y., Varshney, R. K., & Kumar, A. (2022). Mainstreaming orphan millets for advancing climate smart agriculture to secure nutrition and health. *Frontiers in Plant Science, 13*, 902536.
- Choudhary, S., Boruah, A., Ram, N., Gulaiya, S., Choudhary, C. S., & Verma, L. K. (2023). Millet's role in sustainable agriculture: A comprehensive review. *International Journal of Plant & Soil Science, 35*(22), 556-568.
- Eco-Informatics Centre, Conservation Database."Chinnar Wildlife Sanctuary".ATREE.
- Archived fromthe originalon 6 September 2009. Retrieved 2 January 2009.
- Gamble JS, Fischer CEC. (1956). Flora of the presidency of Madras. Vol. Calcutta, India: I– III: Botanical Survey of India.
- Gowri, M. U., & Prabhu, R. (2017). Millet production and its scope for revival in India with special reference to Tamil Nadu. *International Journal of Farm Sciences, 7*(2), 88-93.
- Kheya, S. A., Talukder, S. K., Datta, P., Yeasmin, S., Rashid, M. H., Hasan, A. K., & Islam, A. M. (2023). Millets: The future crops for the tropics-Status, challenges and future prospects. *Heliyon, 9*(11).
- Magdoff, F., & Tokar, B. (2010). *Agriculture and food in crisis: Conflict, resistance, and renewal.* NYU Press.
- Michaelraj, P. S. J., & Shanmugam, A. (2013). A study

on millets cultivation in Karur district of Tamilnadu. *International Journal of Management Research and Reviews*, 3(1), 2167.

- Mrabet, R. (2023). Sustainable agriculture for food and nutritional security. In *Sustainable agriculture and the environment* (pp. 25-90). Academic Press.
- Padulosi, S., Mal, B., King, O. I., &Gotor, E. (2015). Minor millets as a central element for sustainably enhanced incomes, empowerment, and nutrition in rural
- India. *Sustainability*, 7(7), 8904-8933.
- Rao, N. (2007). 2 Global Nutrition Report. Global Scenario of Millets Cultivation, 49(2), 33–45.
- Ramashia, S. E., Mashau, M. E., &Onipe, O. O. (2021). Millet's cereal grains: nutritional composition and utilisation in Sub-Saharan Africa. In *Cereal Grains-Volume 1*. IntechOpen.
- Srivastava, P., Sangeetha, C., Baskar, P., Mondal, K., Bharti, S. D., Singh, B. V., & Agnihotri, N. (2023). Unleashing the potential of millets promoting nutritious grains as vital cereal staples during the international year of millets: a review. *International Journal of Plant & Soil Science*, 35(18), 1860-1871.
- Tripathi, G., Jitendrakumar, P. H., Borah, A., Nath, D., Das, H., Bansal, S., ... & Singh, B. V.
- (2023). A review on nutritional and health benefits of millets. *International Journal of Plant & Soil Science*, 35(19), 1736-1743.

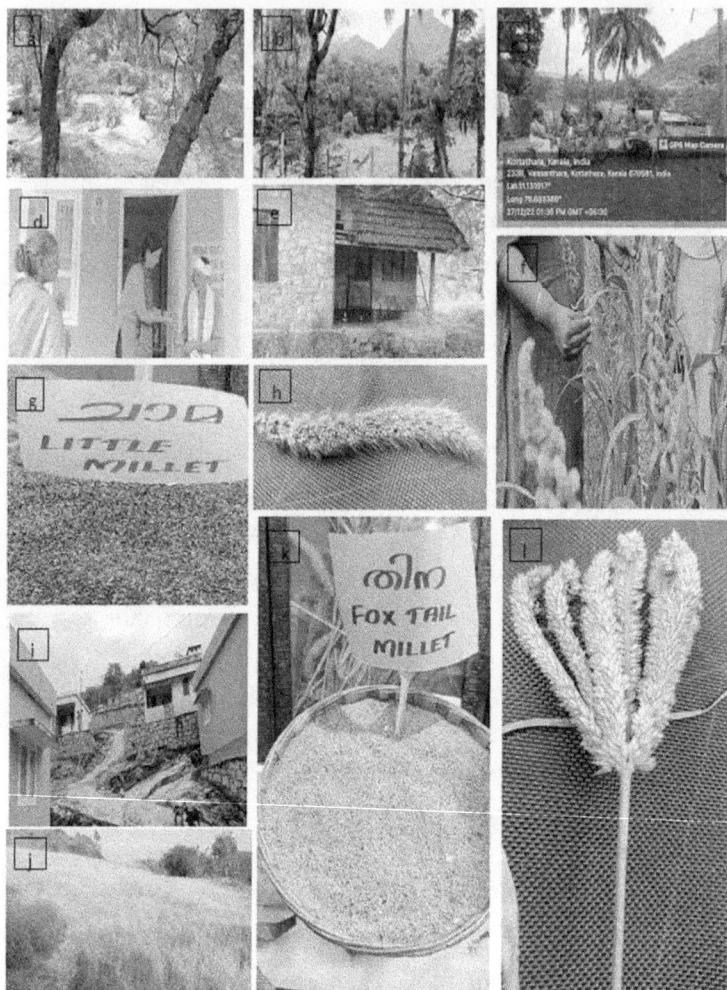

Plate 1: a & b - Chinnar wild life sanctuary, c - Thayyannankudi tribals, d-chieftain Chandran Kani , e-Champakad Ooru, f-kuthiravali ,g -chama millet,h -Thina,i - Thayyannnakudi Ooru, j-Chama, k -Thina, l -Ragi

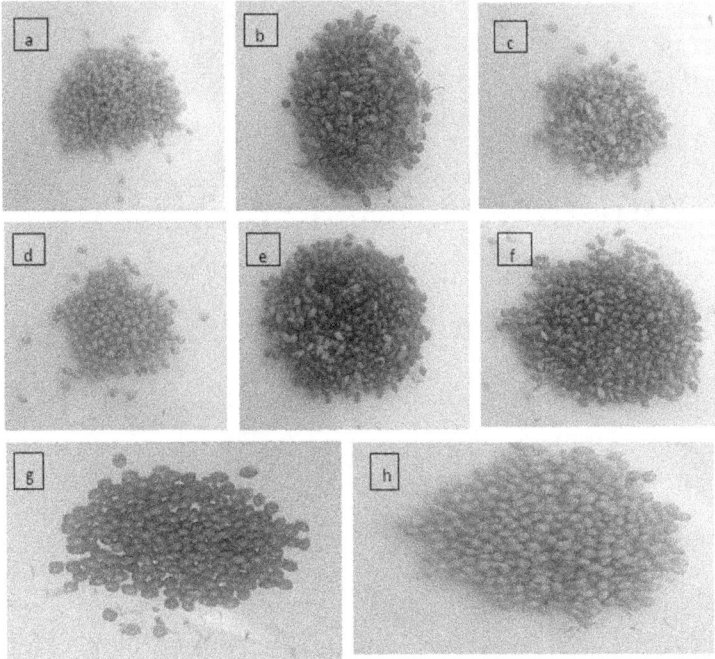

Plate 2: a - Pullu thina, b -Karu sama, c - Vella varagu, d - Periya sama, e - Karu sama, f - Vella sama, g - Mulian thina , h - Kuthira vali

Holistic Healing: Exploring the Synergy Between Ayurveda and Yoga for Wellness

Madhurima MR

Research scholar, Department of English,
All Saints College, Trivandrum

Abstract:

In a fast-paced world people have no time to think about the importance of healthy life and their well-being. People run behind their life dreams and their ambitions. Today people face tension and stress from their field of job and their personal life. In this situation yoga and Ayurveda heals the health issue especially physical health and mental health. It is a holistic healing and helps to balance our mind and body. These two sciences like Ayurveda and Yoga originated from the Indian philosophy. Ayurveda and Yoga are closely interrelated and help to improve our lives. This article discusses the importance of yoga and Ayurveda in day-to-day life.

Key Word: Holistic, Philosophy, Ayurveda, Yoga

In India, the origin of Yoga and Ayurveda dates back 1000 years. It is a holistic approach and helps to decrease the stress of daily life and console the issues faced by the body and mind of a human being. The practice of Ayurvedic promotes a better lifestyle and provides a nutritious diet and Yoga unifies physical and mental health. Ayurveda evolved from the Atharvaveda. The word Ayurveda from the Sanskrit word which means 'science of life'. It is the joining of two words 'Ayu' and 'Veda' Ayu signifies the life of human beings and 'Veda' marks science. The major Goal of Ayurveda is wasthasyaSwasthyaRakshnam Atursasya Vikara Prashaman 'which means, the distribution of better health and the cure of ill health. Ayurveda is a traditional and oldest medicinal practice it has different types of treatments named yoga, massage, acupuncture, herbal medicine etc. Ayurveda believes that everything is related to the Universe and it connects living things, non-living things and the dead. Our human body and mind are connected with the universe and it promotes better health. Sometimes the human body is affected by some issues and it disrupts the harmony of the body and causes illness.

Similarly, a person is built upon fundamental units of the Universe like space, air, fire, water and earth. Based on these, the human body produces three life energies called dosa (Vata dosha, pitta dosha, kapha dosha). Vata Dosha is powerful compared to other dosha and manages the respiratory system, circulatory system, intestine system and mind. Vata Dosha is prominent in the human body because a person should be active, creative and physically the people should be thin. This type of Dosha causes rheumatoid issues, respiratory issues and heart problems. To balance this dosha do meditation, eat hot food etc. Pitta Dosha manages our metabolism and hormone changes. Pitta dosha is dominant in the human body. The person should be a confident and democratic leader. This dose affects the digestion and circulatory system. To balance this Dohsa have vegetables and do yoga. Kapha Dosha controls the muscular System and immune systems. This dosha disrupts breathing systems, sleep etc. For Balancing Kabha dosha use vegetables like carrots, cucumber and do yoga and jogging. To maintain a healthy body practitioners planned Ayurvedic treatment. There are different types of medicine to maintain a healthy body prevent disease and avoid illness.

Some tools for Ayurvedic treatment are Yoga, meditation, herbal medicine, counselling purification programs etc. Meditation maintains our stress and strain and also improves good sleep. Herbal medicine is a combination of ginger, turmeric red clover etc. Counselling is another tool to identify the Dosha and help to maintain the lifestyle of the human being. A purification program means to purify our blood using herbs, oils, massages etc.

Yoga is a traditional medical practice from thousands of years back and the word originated from Rigveda. The word yoga originated from the Sanskrit word 'Yuj', It means Union. Modern yoga popular in the 1970's emphasises the harmony of the whole body. Modern Yoga is defined as Asanas, these Asanas help the well-being of the people. Today the majority of the people are well trained in yoga.

Different types of Yoga, Ashtanga Yoga, Vini Yoga, Restorative Yoga etc Ashtanga Yoga was popularized in 1970 and it helps to improve the breathing system. Vini Yoga is another type of Yoga that helps the function of the breathing. Restorative yoga helps to relax from the stressful mood. Yoga improves breathing ability, and sleep, avoids stress and strain and enhances body strength. Yoga decreases body pain especially knee pain, neck pain etc. Yoga enhances self-care and also cleanses their own body.

Yoga use in our daily life helps physical and mental health. Some Yoga poses are mentioned below, Tadasana, Balasana, Uttanasana, Vrikshasana, Sukhasana, and Bhujangasana.

Integration of Ayurveda and Yoga

These two branches purify our human body and soul. Ayurveda and Yoga originated in India popularised and practised in East and West countries. They have different branches also. It is trying to strengthen our mental, physical, and psychological abilities. Ayurveda and Yoga are balanced with Doshas, Dhatus and Malas. Both of the Sciences use important tools like meditation, prayer, a healthy diet, asanas etc. It improves digestion and other functions of our human body. Both provide balanced help to human beings. The ultimate aim of Ayurveda and Yoga are holistic strength of the human body. In the time of COVID-19 19 Ayurveda and Yoga were highly helpful to the medical field and they prevented disease and improved the immunity power of human beings.

Practical Tips for Choosing Ayurveda and Yoga Lifestyle

- Follow daily routines in lifestyle- Wake up early, eat small quantities and diet food practice yoga and prayer and bed at the correct time.
- Eat food based on your Dosha - maintain diet food and balance Doshas and avoid caffeine, and refined sugar.
- Practice yoga and prayer regularly - Practice Asanas, Prayer, Pranayama etc.
- Importance of Yoga and Ayurveda in our daily life
- Yoga and Ayurveda help to improve the holistic well-being of the human body. Practice regularly improves our physical and mental health.
- Physical Health

Ayurveda helps the human body to maintain a lifestyle and diet based on dosha and helps to follow a natural lifestyle, helping exert the waste from the body. Yoga with the help of Asanas, Meditation, Prayer, and Pranayama improves the strength of the human body and also regulates the immunity system.

Mental and Emotional Health

The practice of Yoga and Ayurveda reduces the stress and strain on a human being and gives peace.

Spiritual Health

Both the use of Yoga and Ayurveda enhances self-realization, meditation, and devotion and it helps the person connect to the universe.

- It helps to improve digestion and our metabolism.
- It decreases the pains in our body, especially knee pain, joint pain, and neck pain.
- It helps to maintain blood pressure and cholesterol levels.
- It helps to maintain proper sleep and well-being.
- It helps to promote self-care.
- Both of the Sciences help to change the skin colour, which means faded skin changes to youthful skin.

Conclusion

Ayurveda and Yoga are philosophically and spiritually connected with our human being. These two transform our human body into comprehensive harmony. Ayurveda and Yoga helps to balance the Doshas in our body. Practising these two in our daily lives helps to promote a happy healthy, stress-free mind. It helps to tackle the emotional problems.

Reference

- Ayur wakeup.'How are Ayurveda and Yoga Related?',25 Jan 2023,https://www.ayurwakeup.com.
- Dabur.'Yoga and Ayurveda Integration: A Holistic System',https:www.dabur.com.
- Devraj,Vignesh.'Yoga and Ayurveda-Importance of Healthy Life'.SitaramRetreat,https://sitaram retreat.com.
- Directorate of Ayush.Ayurveda.2023.https://ayush.delhi.gov.in.
- Dr Athul.'Yoga and Ayurveda Lifestyles for a Healthy Life.DheemahiAyurveda,https://dheemahiayur.com.
- Nichols, Hannah, 'How does Yoga Work?'.Medical News Today,2023,https://www.medicalnewstoday.com.
- The Pulse.'The Yoga-Ayurveda Connect.Maharishi Ayurveda,2021,https://mapi.com.
- Worth, Tommy,' Does it Work?'.WebMD,2023,https://www.webmd.com.

Carbon Nanotubes Properties and Applications: A State-of-the-Art Review

Dr. Sutheertha S. Nair

Department of Physics, VTMNSS College, Dhanuvachapuram,
Thiruvananthapuram- 695503, Kerala, India

Abstract

Carbon nanotubes (CNTs), owing to their inherent one-dimensional nanostructure, exceptional mechanical robustness, and remarkable electronic transport characteristics, have elicited substantial global scientific inquiry. These foundational attributes position CNTs as a uniquely versatile material system with a broad spectrum of high-impact technological applications. This review elucidates exciting properties and critical materials science applications of CNTs, specifically addressing their utilization in advanced electronics, energy storage systems, biosensors, drug delivery systems and space applications. The challenges associated with their commercialization and future perspectives for research are also discussed.

1.   Introduction

The advent of nanotechnology was significantly impacted by the emergence of carbon nanotubes (CNTs).

Their nanoscale dimensions and exceptional properties have positioned them as a cornerstone of the field. Carbon nanotubes are one atom thick graphene sheet rolled in in the form of a cylindrical structure. These hollow cylinders, composed of graphite carbon atoms at the nanometer scale belong to the fullerene family, often featuring hemispherical buckyball. The remarkable versatility of CNTs stems from their size, which dictates a wide spectrum of electronic, thermal, and structural characteristics that can be further tuned by their length, diameter, and chirality. Their high thermal conductivity, Young's modulus, substantial surface area, impressive current density, ability to exhibit ballistic transport at submicron levels, and massless Dirac fermion behavior unlock their potential in diverse applications, including nanoelectronics, reinforced materials, bio sensors, drug delivery, water filtration, space applications etc (1-5). Because of their singular characteristics, including their production, purification, and various applications, CNTs have become a significant and absorbing area of exploration in nanotechnology. This review offers a general perspective on the application of carbon nanotubes in composite materials and the challenges ahead for them, which furnishes the necessary details for continued research.

2. Types of Carbon Nanotubes

Carbon nanotubes are one atom thick graphene sheet rolled in to a cylindrical structure. Based on the way in which the graphene sheets are rolled up or twist in the tube structure, described by chirality, carbon nanotubes are again classified in to armchair, zigzag and chiral nanotubes. Figure 1represents the different orientation patterns of carbon nanotubes

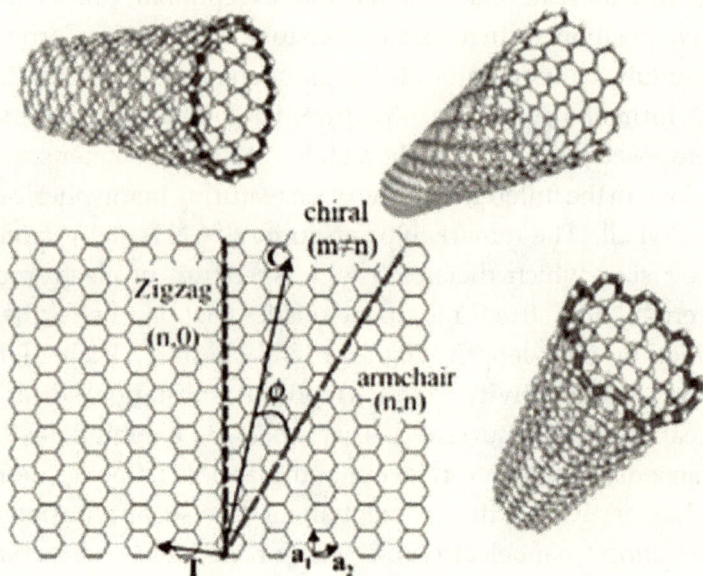

Figure 1. Different orientation patterns of carbon nanotubes (ref S. Abdalla etal.

In armchair nanotubes the carbon atoms form a repeating zigzag pattern along the tube's circumference, resembling the arms of an armchair and in zigzag nanotubes the carbon atoms form a zigzag pattern along the tube's longitudinal axis. In chiral nanotubes the graphene sheet is rolled up at an angle, resulting in a helical pattern of carbon atoms.

Based on the number of graphene layers in their structure carbon nanotubes (CNTs) are classified in to two types, Single-walled Carbon Nanotubes (SWCNTs)and Multi-walled Carbon Nanotubes (MWCNTs). Figure 2 represents single walled and multiwalled carbon nanotubes.

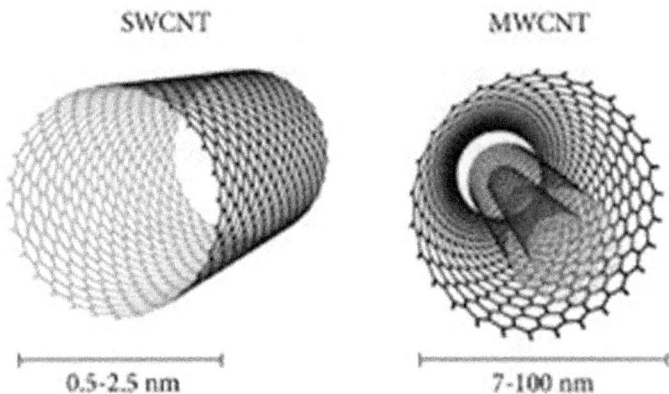

SWCNT   MWCNT

0.5-2.5 nm   7-100 nm

Figure 2. Single walled and multi walled carbon nanotubes

Single-walled Carbon Nanotubes (SWCNTs) consist of a single layer of graphene sheet rolled up into a seamless cylinder and Multi-walled Carbon Nanotubes (MWCNTs) composed of multiple concentric layers of graphene sheets arranged in a nested fashion. Multiwalled carbon nanotubes generally have larger diameters compared to SWCNTs, ranging from 5-50 nanometers.

3. Properties of carbon nanotubes

CNTs are one of the strongest materials ever discovered with exceptional mechanical strength, having tensile strength many times greater than steel. This makes them ideal for use in composite materials that can be used to make lighter and stronger structures. Electrical conductivity of carbon nanotube is very high so that they have a wide variety of electronic applications, such as transistors, sensors, and solar cells. Excellent thermal conductivity of the carbon nanotubes makes them potential candidate for heat sinks and other thermal management applications. Due to large surface area to volume ratio carbon nanotubes are used in applications such as catalysis, where they can be used to speed up chemical reactions. Due to strong covalent bonds connecting each carbon atom, they have high melting point which contributes to their efficient electron emission capabilities. Surprisingly, CNTs exhibit a significant degree of elasticity, allowing them to elongate up to 18% before failure.

4. Synthesis of carbon nanotubes

There are several methods for synthesizing carbon nanotubes (CNTs), The major techniques used for the synthesis of carbon nanotubes are

4.1 Arc Discharge:

This is the oldest and simplest method, involving an electric current between two graphite rods in a vacuum or an inert gas environment. This will create high-temperature plasma that vaporizes carbon, which then self-assembles into nanotubes on the cooler cathode and produces high-quality carbon nanotubes.

4.2 Laser Ablation:

In laser ablation method a pulsed laser vaporizes a graphite target in a high-temperature reactor. The vaporized carbon condenses on cooler surfaces to form carbon nanotubes. Laser ablation method offers more control over diameter of the nanotubes than arc discharge method, but production yield is lower. Fig 3 represents image of laser ablation technique for the synthesis of nanoparticle

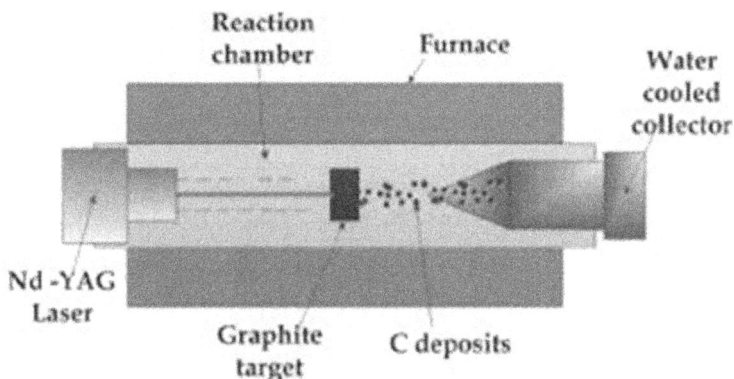

Fig 3 laser ablation technique for the synthesis of nanoparticle

4.3 Chemical Vapor Deposition (CVD):

This is the most promising method for large-scale production of carbon nanotubes. This is a process of decomposing a gas containing carbon atoms (like hydrocarbons) into smaller components, including carbon atoms, which then deposit on a substrate to form a solid material. Using chemical vapor deposition properties like diameter, wall structure and alignment of carbon nanotubes can be controlled.

4.4 High-pressure carbon monoxide disproportionation (HiPCO) and Plasma-Enhanced Chemical vapor Deposition (PCVD):

HiPCO uses a high-pressure, high-temperature carbon monoxide reaction with a metal carbonyl catalyst, and plasma-enhanced CVD using plasma to facilitate chemical vapor deposition.

4.5 Flame synthesis:

Flame synthesis technique is used for the synthesis of carbon nanotubes using acetylene jet flames and galvanized steel catalyst. It was found that acetylene flames generate coiled nanotubes within a narrow pyrolysis zone above the burner for a limited range of fuel flow rates

5. Applications of carbon nanotubes

Carbon nanotubes (CNTs) hold immense promise due to their unique combination of properties. The extraordinary properties of carbon nanotubes (CNTs) make them prime candidates for a variety of applications. The potential applications of CNTs are constantly being explored as research progresses. Some of the applications of carbon nanotubes in various fields are discussing below (1-5).

5.1 Electronics:

Carbon nanotubes are strong candidates for next-generation transistors due to their exceptional electrical conductivity and small size, enabling faster and more efficient electronic devices. Their high sensitivity and ability to detect a wide range of molecules make them ideal for chemical and biological sensors. The miniaturization capabilities of CNTs open doors for creating ultra-small electronic components for advanced devices in nanoelectronics.

Field emission stands out as a highly desirable electron source when compared to thermionic emission. Consequently, electron field emission materials have been rigorously investigated for their implementation in advanced technologies such as flat panel displays, electron guns for electron microscopy, and microwave amplification. The specific blend of attributes found in carbon nanotubes including their diameter at the nanometer level, structural soundness, high electrical conductivity, and chemical inertness renders them well-suited as effective electron emitters. In contrast to traditional emitters, carbon nanotubes demonstrate a reduced turn-on electric field. The current-handling capacity and emission consistency among different carbon nanotubes show significant variations based on their manufacturing method and growth parameters

5.2 Reinforcement materials:

The outstanding strength-to-weight ratio of CNTs makes them valuable for reinforcing composites used in aircraft, automobiles, and sporting goods, leading to lighter and stronger structures. CNTs can be incorporated into polymers to create materials with both electrical conductivity and the desirable properties of polymers, like flexibility (2).

5.3 Energy:

CNTs can be used to improve the capacity and lifespan of lithium-ion batteries, leading to longe lasting and more efficient energy storage solutions. CNTs can enhance the efficiency of solar cells by improving light absorption and electron transport. Their large surface area makes them potentially useful for storing hydrogen fuel, a clean energy source. Carbon nanotubes are being evaluated for their use in energy production and storage. In contrast to graphite, carbon-based materials, and carbon fiber electrodes, which have a long history in fuel cells and batteries, nanotubes offer unique benefits. Their small size, smooth surface, and specific surface are advantageous. The speed of electron transfer at carbon electrodes is crucial for fuel cell efficiency. It was reported that the electron transfer mechanism in carbon nanotubes is faster compared to conventional carbon electrodes.

5.4 Drug delivery:

CNTs can be used as carriers for targeted drug delivery, allowing for precise medication release within the body. Their elongated structure allows them to carry a large payload of drugs compared to their size. By functionalizing CNTs (attaching specific molecules to their surface), scientists can control their solubility, targeting ability, and release behavior of the encapsulated drug. CNTs can potentially penetrate cell membranes, delivering drugs directly into cells, which can be particularly effective for targeted therapies. While some research is ongoing, certain types of CNTs exhibit good biocompatibility, meaning they are not harmful to living tissues.

Drugs can be physically loaded within the hollow core of CNTs and can be chemically attached to the outer surface of CNTs thus allowing a controlled release of the drug at the target site. CNTs can be functionalized to target specific cancer cells, delivering chemotherapy drugs directly to the tumor site. This can minimize side effects on healthy tissues. CNTs may be used to deliver drugs for diseases like Alzheimer's or Parkinson's by crossing the blood-brain barrier, a major hurdle in treating these conditions. CNTs can potentially be used as carriers for gene therapy, delivering genetic material to targeted cells for treatment of various diseases.

5.5 Biosensors:

Similar to regular sensors, CNT-based biosensors can detect biological molecules like proteins or DNA, aiding in medical diagnosis. Enzyme biosensors are the most favored type of biosensor, and many commercial biosensing devices are built upon them. The large surface area of CNTs, which is advantageous for biomolecular attachment, has been utilized in the creation of enzyme biosensors.

5.6 Water filtration:

CNT membranes can effectively remove contaminants from water due to their small pore size and filtering capabilities. Depending on their application, CNT membranes are broadly classified into two categories: (1) Self-supporting CNT membranes and (2) CNT nanocomposite membranes. Self-supporting CNT membranes, which often take the form of vertically aligned CNT (VACNT) membranes, are employed in areas such as desalination and water treatment. In VACNT the cylindrical pores allow the fluid to pass through the spaces between the CNT bundles. All these membranes are highly effective owing to the surface area and size of their constituent nanotubes

5.7 Space

CNT composites used in spacecraft manufacturing can significantly reduce the weight of spacecraft. This translates to lower launch costs and the ability to carry more scientific equipment or cargo. Current space electronics are vulnerable to damage from cosmic rays and solar radiation. CNTs show promise in creating more radiation-resistant transistors and other electronic components, leading to more reliable spacecraft operations. CNTs are excel in conducting heat. This property can be utilized in heat sinks for electronics or even spacesuit temperature regulation systems, ensuring optimal operating temperatures in space. The exceptional electrical conductivity of CNTs can lead to the development of more efficient and powerful antennas for communication and data transmission in space. CNT-based sensors can be highly sensitive and miniaturized, making them suitable for detecting various environmental factors or biomarkers in space missions. The theoretical "space elevator" concept proposes a super-strong tether reaching from Earth's surface to geosynchronous orbit, enabling easier access to space. CNTs, with their exceptional strength-to-weight ratio, are a potential candidate material for such a tether, though significant technological advancements are needed

6.   Challenges and Future Perspectives

While the unique properties of CNTs position them as transformative materials for a multitude of applications, overcoming the economic, structural control, and safety challenges is critical. Continued interdisciplinary research and development focused on scalable, precise, and responsible production and utilization strategies will pave the way for the widespread integration of CNTs into commercial products and technologies.

7.   Conclusion

This review has presented properties and different characteristics and a range of potential applications for carbon nanotubes, particularly highlighting materials science-based applications. The remarkable versatility and predictable properties of carbon nanotubes arising from its well-defined crystal lattice, are the reasons for the considerable excitement in the field of nanotechnology. As a true bridge between the nano and macro worlds, carbon nanotubes are undoubtedly destined for a leading role in future technology.

References

- M.Kim, D.Goerzen, P. V. Jena, E. Zeng, M.Pasquali,R A. Meidl, (2024)nat. rev. mater.9, 63
- P. M. Ajayan1 and O. Z. Zhou Carbon nanotubes:

synthesis, structure, properties, and applications Appl. Phys. (2001)80, 391

- S. Abdalla, F. Al-Marzouki, Ahmed A. Al-Ghamdi and A. Abdel-Daiem Nano.
- Res. Lett (2015) 10:358
- M.Endo1, T. Hayashi1, Y. Ahm Kim1, M. Terrones and M. Dresselhaus (2004) Phil. Trans. R. Soc. Lond. A 1437
- N. Yanga, X. Chena, T.Renb, P. Zhang, D.Yanga Sensors and Actuators B(2015) 207, 690

Nature and Culture: The Vital Role of the Environment in Ancient India

Dr Raji Chandra M

Assistant Professor, PG Dept of History, VTM NSS COLLEGE, Dhanuvachapuram

The geography of India had a very important role in its history. Due to the Himalayas in the north and western ghats and eastern ghats isolated India in Asia. And it helped India for individualised life and development. As we all are well familiar with the environment, it is everything which surrounds us naturally. "**Paryavaranam**" is a Sanskrit word for environment that was prevalent in ancient India, thousands of years before the advent of modern science. A new approaches to environmental protection and conservation are needed and that must be based on interconnectedness, interrelatedness and interdependency of the various components of the earth. The task of protecting the environment and keeping the ecology balanced and safe is the duty of every individual. Our Sanskrit Literature have scientific and religious description of nature as well as conservation of idea about indigenous system of restoration and stablisation.

So it is clear that the matter of preservation and

protection of the environment is not only confined to the contemporary India but also goes way back in history proving its significance.Ancient Indian literature serves as vast reservoirs of knowledge related to everything about environment. The Vedic, Puranic, Jain and Buddhist traditions deals with the principles of ecological harmony centuries ago.

Introduction

India has always had a rich ancient tradition of protecting the environment which in turn, has made the people of India worship and embrace nature in every way possible. Trees, water, animals, land have an important mention in ancient Indian texts. Indian texts such as the Arthashastra, Sathapatha Brahmanas, Vedas, Manusmriti, Ramayana, Mahabharata etc. enable us to understand the concepts of environment conservation and maintaining forest ecology; also hymns in the four Vedas, Rigveda, Yajurveda, Samaveda, and Atharvaveda, reveal full cognizance of the undesirable effects of climate change, distortion in ecological balance, and environmental degradation; and appropriately caution against them.

The Role of Environment in Ancient India

The environment played a crucial role in shaping the civilization, culture, and economy of ancient India. The natural surroundings influenced religious beliefs, agricultural practices, settlement patterns, and the overall way of life of people in that era. From the fertile plains of the Indus Valley to the dense forests mentioned in ancient texts, nature was deeply revered and formed the foundation of Indian civilization.

Vedic Texts (Rigveda, Atharvaveda, Upanishads) –contain hymns and philosophies emphasizing nature's divinity and the need for ecological balance. The Vedas attach great importance to environmental protection and purity. they persist on safeguarding the habitation, proper afforestation and non-pollution. in fact, man is forbidden from exploiting nature. he is taught to live in harmony with nature and recognize that divinity prevails in all elements, including plants and animals. The rishis of the past have always had a great respect for nature. a verse from Rig-Veda says, "thousands and hundreds of years if you want to enjoy the fruits and happiness of life then take up systematic planting of trees."

Arthashastra by Kautilya – discusses policies on forest management, irrigation, and conservation strategies. This treatise provides lot of knowledge about environment and its conservation. it describes the maintenance of public sanitation and preservation of environment, forest and wildlife. even in the affairs of the state, the administration and the ruler were directed to preserve and promote environmental welfare. in the arthasastra, kautilya suggests the need to develop abhayāranya or abhayavana, forest and animal sanctuaries.

Manusmriti – Provides guidelines on environmental ethics, including forest conservation and sustainable resource use.

Jain and Buddhist Scriptures – Promote the principle of Ahimsa (non-violence), advocating for the protection of all living beings, including plants and animals.

Indus Valley Civilization Archaeological Findings – Showcases advanced urban planning and water conservation techniques.

Ancient Indian Temple Inscriptions and Rock

Edicts (e.g., Ashoka's Edicts) – Highlight environmental conservation policies implemented by rulers.

Puranas (e.g., Matsya Purana, Vishnu Purana) – Narrate stories emphasizing the importance of rivers, trees, and sacred groves in Indian tradition.

Bhagavat Gita : Lord Krishna Says InThe Bhagavad Gita (9.26):*"Patram PushpamPhalam*

Toyam, Yo Mey BhaktyaPrayachchatiTadaham Bhakt YupahrutamAsnaami

Prayataatmanaha" I Accept A Leaf, Flower, Fruit Or Water Or Whatever Is Offered With Devotion.

Environment's Influence on Religion and Philosophy

Ancient Indian culture was deeply intertwined with nature. Many religious beliefs and practices were rooted in environmental elements. Rivers like the Ganges, Yamuna, and Sarasvati were considered sacred and worshiped as divine entities. The Rigveda, one of the oldest texts, contains numerous hymns praising nature, including the sun, wind, and rain. The Upanishads and other philosophical texts emphasize harmony between humans and nature, advocating for sustainable living and respect for all life forms.The Hindu Religion Also Stresses Awareness InThe Conservation Of Trees.

The worship of trees and animals was a significant aspect of ancient Indian traditions. Sacred groves were preserved for religious and ecological purposes. Trees like the peepal and banyan were venerated, and animals such as cows, snakes, and elephants held symbolic significance. This reverence for nature helped in conservation efforts and maintaining ecological balance.

Impact on Agriculture and Economy

The agrarian economy of ancient India was heavily dependent on environmental factors. The monsoon played a vital role in determining agricultural productivity. The Indus Valley Civilization (c. 2600–1900 BCE) thrived due to the fertile lands and well-developed irrigation systems. People adapted their farming techniques based on seasonal changes and river patterns, ensuring sustainable agricultural practices.The use of organic farming methods, crop rotation, and natural fertilizers was prevalent, indicating an understanding of ecological balance. Deforestation was generally avoided, and forests were managed with care to prevent soil erosion and maintain biodiversity.

Settlement Pattern

Environmental conditions greatly influenced the location and development of ancient Indian settlements. The Indus Valley Civilization, with cities like Mohenjo-Daro and Harappa, demonstrated advanced urban planning techniques that took into account water conservation, sanitation, and sustainable infrastructure. The presence of granaries, drainage systems, and water reservoirs indicates a strong awareness of environmental factors. Many settlements were built near rivers and fertile plains, ensuring easy access to water for drinking, irrigation, and trade. The forests provided timber, medicinal herbs, and other resources essential for daily life.

Ethics and Conservation

Ancient India demonstrated a deep respect for the environment, viewing nature as sacred and integral to life, with practices like protecting forests, venerating animals, and emphasizing water conservation, as seen in texts like the Artha Shastra and religious traditions.

Ancient Indian texts promote the idea of environmental conservation. The Arthashastra, written by Kautilya, mentions forest management, irrigation policies, and wildlife protection. The Manusmriti and other dharma texts lay down rules for the protection of natural resources, emphasizing the importance of maintaining ecological balance. Kings and rulers implemented policies to protect forests, preserve water bodies, and maintain biodiversity. Certain areas were designated as protected spaces where hunting and deforestation were restricted. The philosophy of non-violence (Ahimsa), particularly in Jainism and Buddhism, extended to environmental ethics, advocating for the protection of all living beings.

Here's a more detailed look at the environmental awareness and practices in ancient India:

Sacredness of Nature: Ancient Indians viewed nature, including mountains, rivers, forests, and animals, as sacred and deserving of respect and protection.

Mother Earth Concept: The concept of "Mata Bhoomi" (Mother Earth) and "Pithru Bhoomi" (Father Earth) reflected a deep connection to the land and its resources.

Vedic Texts: Vedic scriptures emphasized the importance of protecting forests and wildlife, with cutting green trees being considered a punishable offense.

Religious Traditions: Various religious traditions, including Hinduism, Buddhism, and Jainism, promoted environmental ethics, with Jainism's principle of "Ahimsa" (non-violence) extending to all living beings.

Forest Gods and Goddesses: Forests were associated with deities, fostering a sense of reverence and protection.

Environmental Practices:

Water Management: Ancient Indian civilizations,

particularly the Indus Valley Civilization, were known for their advanced water management systems, including wells, public baths, and covered underground drains.

Sanitation: The Indus Valley Civilization had a strong focus on hygiene and sanitation, with evidence of ventilated houses, orderly streets, and well-maintained water systems.

Forest Management: The Artha Shastra, a treatise on economics and statecraft, provided guidelines for managing forests, emphasizing their preservation for future generations.

• **Wildlife Protection:**Ashoka, a Mauryan emperor, banned hunting certain species of wild animals and established forest and wildlife reserves.

• **Traditional Techniques:**Villages often had sacred lakes and groves of trees to catch rainfall, protect banks from erosion, and provide water for irrigation and drinking.

• **Planting Trees:**Planting trees was considered a virtuous act, with some texts mentioning that planting certain trees could lead to positive outcomes in the afterlife.

For example :

The Mahabharata mentions the importance of trees, stating that "Trees with flowers and fruits fulfil this earth". The Manusmriti emphasizes the importance of protecting the environment and prohibits polluting water sources. The Puranas like Varaha Purana states that planting certain trees and plants can protect from falling into hell. Kautilya's Artha Shastra contains comprehensive guidelines for managing forests, with a focus on their preservation for future generations.

Modern Relevance:

1. **Lessons for Sustainability:**The environmental practices and ethics of ancient India offer valuable

lessons for modernday sustainability efforts.

2. **Need for Revival:**Some argue that modern society can learn from the environmental wisdom of ancient India and adopt traditional practices to address contemporary environmental challenges.

3. **Environmental Awareness:**The concept of "Ahimsa" and the reverence for nature in ancient Indian traditions can help foster a greater sense of environmental awareness and responsibility.

As per ancient Hindu philosophy, everything in this universe operates as per the supreme law of the universe called „Rita⊚. There does not seem to be a term in today⊚s context which is equivalent to the term „Rita⊚.

Conclusion

The environment was an integral part of life in ancient India, influencing religious beliefs, agriculture, urban planning, and ethical practices. The deep respect for nature helped in conservation and sustainable living. By revisiting these ancient principles, we can find solutions to modern environmental challenges and promote a harmonious relationship between humanity and nature.

While ancient India had strong environmental ethics, challenges like deforestation, excessive urbanization, and soil depletion occasionally arose. Overuse of natural resources sometimes led to ecological imbalances, contributing to the decline of some civilizations.

Modern society can draw valuable lessons from ancient India's environmental practices. Sustainable agriculture, water conservation, forest preservation, and respect for biodiversity are principles that remain relevant today. Reviving traditional knowledge and integrating it with modern environmental policies can help address contemporary ecological crises.

References

- B Shyamala, Relevance of Ancient Indian Methods of Environmental Protection in the
- Present Day Scenario, International Review of Business and Economics, 2018
- D.B.N. Murthy, Environmental Awareness & Protection: A Basic Book on Evs. New Delhi: Deep and Deep publicationsPvt. Ltd.,2004.
- Dr Ren Tanvar(art), IOSR-JHSSISSN-2279-0837.,2016
- Http:///Hinduwisdom.Info/
- K Ratnabali (art), Rethinking Approach to Environmental Protection in view of Ancient Indian Wisdom, Scholars international School of Law crime and Justice, UAE, 2020.
- Mukundananda, S. (2014). Chapter 3, Verse 10 – Bhagavad Gita, The Song of God – Swami Mukundananda. Retrieved November 14, 2020, from https://www.holybhagavadgita.org/chapter/3/verse/10.
- Panikkar, K M., A Survey of Indian History, LG Publishers , Delhi, 1947.
- Rajani Rao U, (art), International journal of Life Science Research, ISSN 2348-313X, 2014
- R. Shamasastry, (Tr), Kautilya⊚sArthashastra< http://educonnectu.yolasite.com/resources/Arthashastra-of- Chanakya-English.pdf.
- *http://www.sanskritimagazine.com/indianreligions/hinduism/nature-worship/*

How Social Media Westernises Indian Indigenous Knowledge Systems: A Media Theoretical Perspective

Abijith P S

University Of Kerala
Email: abijithps1399@gmail.com
Phone: +91 83048 29822

Abstract

This study looks at how cultural reframing and selective depiction on social media can westernise Indian Indigenous Knowledge Systems (IKS). Traditional disciplines like Ayurveda, yoga, and artisanal crafts are increasingly being presented in Westernised formats that put acsthetics and marketability ahead of authenticity, while platforms like Instagram, TikTok, and YouTube dominate public opinion. The paper examines how repeated exposure, influencer behaviour, algorithm-driven visibility, and public silence about marginalised narratives cause IKS to be distorted, drawing on theories such as Framing Theory, Cultivation Theory, Social Learning Theory, and the Spiral of Silence Theory. In the era of algorithmic media, this article emphasises the critical necessity for deliberate digital participation, the promotion of indigenous voices, and the preservation of cultural integrity by examining instances of cultural appropriation and content trends. IKS runs a serious risk of losing its philosophical depth and historical validity in international discourse if proactive measures are not taken to recover and contextualise it.

Keywords: Indian Knowledge Systems (IKS), Social Media, Westernisation, Cultural Appropriation, Media Theory.

Introduction

Social media has revolutionised the exchange and use of knowledge by making it widely available and instantly accessible. In contrast to conventional methods of communication, social media platforms enable users to interact with online content in real-time, frequently making it difficult to distinguish between trend, opinion, and truth. In addition to providing information, this continuous stream of information also influences public understanding through supporting some narratives while repressing others. Social media is an effective tool in creating modern cultural and social realities because users' perceptions, beliefs, and understanding of the world are gradually shaped by similar content that they browse.

In addition to encouraging trends, social media has a significant impact on public opinion by progressively changing how people interact with culture. An example of this change is the rise of de-influencing, in which people question prevailing or commercial narratives and discourage mindless devotion to popular products. Increasing consciousness of the audience is reflected as they start to wonder about the sincerity and intentions behind products, lifestyles, and cultural representations they consume. Social media greatly facilitates this process by amplifying counter-narratives and legitimising changes in perception, as noted by Elhajjar and Itani (2025), demonstrating how online platforms can transform cultural understanding and values.

The #deinfluencing hashtag, with an impressive 1.5 billion views (as of March 2024), garnered most of these views within one year. Further, there were over 28,400 videos under the #deinfluencing hashtag on TikTok, many of which are trending and amassing millions of views each. (Elhajjar and Itani 2)

Social Media has been a popular platform for sharing Indian Indigenous Knowledge Systems (IKS), which include Ayurveda, yoga, traditional art, and folk medicine. But this exposure has also resulted in their Westernisation, when historical authenticity and cultural nuance are frequently sacrificed for global marketability. Research shows that social media affects how people see culture. According to a Pew Research Centre study from 2021, 64% of social media users worldwide think that platforms affect how people perceive history and culture(Auxier and Anderson). This research explores how social media influences the perception of IKS, leading to a shift from indigenous traditions to Westernised interpretations.

The term "Indian Knowledge Systems" (IKS) describes the ancient, varied, and rich traditions of knowledge that evolved over thousands of years in the Indian subcontinent. These systems cover a wide range of disciplines, including linguistics, medicine, science, philosophy, art, architecture, agriculture, and environmental practices. IKS is a holistic perspective that incorporates body, mind, nature, and spirit, and it is based on indigenous wisdom and cultural values. The importance of IKS has been increasingly acknowledged in recent years, both in India and abroad, particularly in traditional sciences, mental wellness, and sustainability.

Indian knowledge systems include a wide range of topics and provide perspectives on nearly all aspects of human life. Ayurveda and Siddha offer an integrated approach to healing in the field of medicine that takes account of the environment, body, mind, and spirit. These approaches prioritise harmony, balance, and prevention over the discrete management of symptoms. Philosophical traditions that explore intricate concepts of awareness, logic, and metaphysics include Vedanta, Samkhya, and Nyaya. These traditions provide conceptions of reality that transcend materialist frameworks. Astronomy and mathematics are other important subjects in IKS. The decimal system, advanced trigonometry, and zero were all introduced by ancient Indian intellectuals and were essential to the growth of modern mathematics. Astronomical literature like *Aryabhatiya* and *Surya Siddhanta* shows a sophisticated comprehension of eclipses, planetary motions, and timekeeping systems. Modern linguistic philosophy has been affected by Indian linguistic heritage, particularly Panini's grammatical book *Ashtadhyayi*, which is regarded as a foundational work in formal language analysis. IKS promotes ecological and agricultural sustainable methods that put the needs of the natural world first. Stepwells and tanks are examples of indigenous methods of water conservation that were created to manage limited resources. Seasonal agricultural cycles, organic farming practices, and sacred groves all demonstrate an innate awareness of ecological balance. These systems viewed nature as a living thing to coexist with rather than as a resource to be exploited. IKS is also strongly influenced by Indian aesthetics and the arts. Texts like the *Natyashastra*, which describe the fundamentals of rhythm, emotion, and expression, serve as a guide for traditional dance forms, music, drama, and architecture. These are regarded as spiritual acts that aim to raise human consciousness rather than just being an amusement.

Indigenous knowledge systems were ignored when colonisation arrived. Traditional Indian methods were frequently characterised as archaic or superstitious, while Western educational and scientific paradigms were imposed as superior. A large portion of IKS was thereby marginalised and lost institutional recognition. Local communities lost the ability to freely practise native medicine, manage their resources, and educate their children in traditional ways as a result of colonial policy. A generational divide in the spread of IKS resulted from this disturbance. Indigenous practitioners were frequently left out of official systems of knowledge production, oral traditions deteriorated, and classical texts lost value. Nevertheless, despite systematic suppression, resistance persisted in the shape of local literature, folk customs, and community-based instruction that subtly preserved certain aspects of IKS.

Interest in Indian knowledge systems has gradually increased in India since independence. IKS has been reintroduced into academic curricula by organisations such as the Ministry of Education and several research councils. The significance of incorporating traditional knowledge into contemporary education is emphasised in the National Education Policy (NEP) 2020. The growing worldwide appeal of yoga, meditation, Ayurveda, and Sanskrit studies is another indication of this renaissance.

The public's opinion of IKS has always been greatly influenced by the media. IKS was frequently framed, constrained or simplified by traditional media, including television, newspapers, and documentaries. It was either presented as mystical and spiritual, appealing to foreign curiosity, or as antiquated and outdated, supporting colonial notions of intellectual supremacy. How seriously IKS was regarded in the fields of academia, medicine, and policymaking was influenced by these representations. Because of this, traditional systems like Ayurveda, Yoga, or Vastu were frequently romanticised or disregarded instead of being comprehended in their original scientific and philosophical contexts.

The way IKS is portrayed has changed significantly since social media's introduction. Individual practitioners, academics, and cultural lovers can now communicate indigenous knowledge with audiences directly, thanks to platforms like Instagram, YouTube, and TikTok, which have increased visibility and involvement. But there are risks as well as opportunities associated with this democratisation. Traditional methods have been diluted and distorted, even though it has made IKS more approachable and rekindled interest among new generations. Commercial businesses and influencers frequently reduce complicated systems to consumable trends, transforming religious rituals into beautiful content, Ayurveda into spa treatments, or yoga into a fitness regimen. The underlying connotations, cultural context and ethical frameworks of IKS are sometimes overlooked in the rush for marketability and virality.

IKS has become much more westernised as a result of this trend. Many aspects of IKS have been repackaged to accommodate the tastes of global consumers, despite having their roots in Indian metaphysics, communal traditions, and spiritual ideologies. Western validation serves as a criterion for legitimacy, whereby old knowledge is only seen as legitimate if supported by international organisations or scientific research. People's perspectives, values, and interactions with their heritage are altered by this transition. Cultural distortion is not the only threat; the voices, languages, and communities that have supported IKS for many generations could also be erased. Promoting critical participation and culturally grounded narratives that uphold the integrity of Indian Knowledge Systems is crucial as social media continues to influence public opinion.

To determine how social media contributes to the westernisation of Indian knowledge systems, this study will employ the framing theory, cultivation theory, social learning theory, and spiral of silence theories. The study aims to identify the fundamental mechanisms that modify public perception and progressively match indigenous knowledge with Western norms and aesthetics by examining how traditional activities are depicted, constantly reinforced, replicated, or repressed across media channels. When combined, these ideas offer a thorough framework for comprehending how media actively creates cultural change in addition to reflecting it.

Indian Indigenous Knowledge Systems are frequently rebranded to align with Western wellness trends. Ayurveda is often portrayed as a holistic lifestyle rather than a traditional medicinal system, and yoga is mainly depicted as a fitness practice rather than a spiritual discipline. Instagram influencers and brands often reframe these practices to appeal to Western audiences, overshadowing authentic Indian practitioners. Western brands often rebrand Indian artisanal traditions as trendy products, overshadowing authentic sources. (Askegaard and Eckhardt 50). This phenomenon can be understood through framing theory, which explains how the media selectively presents information to shape public perception. Goffman's foundational text, *Frame Analysis,* introduced the concept of "frames" as mental structures that shape the way individuals perceive and interpret events and experiences in social life. A study by Entman (1993) highlights that media framing influences how audiences interpret cultural narratives. Entman says,

Framing essentially involves selection and salience. To frame is to select some aspects of a perceived reality and make them more salient in a communicating text in such a way as to promote a particular problem definition, causal interpretation, moral evaluation, and/or treatment recommendation for the item described. (52)

Framing, when used in media studies, describes how information is presented to audiences to influence their perceptions and reactions. Media frames affect public opinion and conversation by emphasising some parts of a narrative while leaving out others. A limited and occasionally skewed knowledge results, for instance, when ancient practices like yoga or ayurveda are presented only as wellness fads, ignoring their underlying cultural and philosophical value. For instance, the hashtag #GoldenMilk has gained significant traction among Western influencers, presenting turmeric milk as a trendy wellness drink rather than its traditional Indian use as "haldidoodh," a centuries-old remedy for colds and inflammation. The repeated exposure to Westernised representations of IKS influences audience perceptions, making them accept these altered versions as the norm. Western influencers promoting turmeric-based wellness drinks garner more engagement than Indian Ayurvedic experts discussing the same ingredient in its traditional context. Audience views are influenced by prolonged exposure to Westernised depictions of IKS, leading them to accept these modified versions as the standard. Influencers from the West who promote wellness drinks made with turmeric receive more interaction than Indian Ayurvedic specialists who talk about the same substance in its traditional context.

George Gerbner's Cultivation Theory investigates how people's views of social reality are progressively shaped by extended exposure to the content pushed out by the media. The theory says that mass media, including television, provide stories that gradually create a common worldview. Constant exposure to Western-centric media portrayals of Indian Knowledge Systems (IKS) contributes to the belief that indigenous traditions are outmoded or subpar and that Western science, lives, and values are the standard. IKS is marginalised and Western alternatives are preferred as a result of this trained worldview's subtle influence on attitudes and beliefs. The study tries to state how the media persuades the viewer to form an assumption based on the constructed reality:

The dominant stylistic convention of Western narrative art-novels, plays, films, and TV dramas is that of representational realism. However contrived television plots are, viewers assume that they take place against a backdrop of the real world. Nothing impeaches the basic "reality" of the world of television drama. It is also highly informative. That is, it offers to the unsuspecting viewer a continuous stream of "facts" and impressions about the way of the world, about the constancies and vagaries of human nature, and the consequences of actions. The premise of realism is a Trojan horse which carries within it a highly selective, synthetic, and purposeful image of the facts of life. (Gerbner and Gross 178)

Cultivation Theory explains how sustained media

exposure shapes long-term beliefs, gradually shifting public understanding of IKS toward Western narratives. A study by Dennis, Alan R., Antino Kim, and Brittany Marett on fake news on social media Dennis, Alan R., Antino Kim, and Brittany Marett found that "Participants were more likely to believe and process headlines that aligned with their beliefs. This likely did not result from additional knowledge; participants were not correct in their perceptions of truth"(21). This is evident in the popularity of influencers like @drmarkhyman, who frequently discuss Ayurveda-inspired health remedies but without reference to their cultural roots, leading audiences to associate these practices more with modern health trends than with their original indigenous contexts. Users of social media, especially younger audiences, often copy what they see on social media. Many Indian influencers emulate Western health and wellness pioneers who, while ignoring the historic relevance of Ayurvedic and yoga techniques, popularise them with commercialised aesthetics.

According to Social Learning Theory, individuals adopt behaviours by observing and imitating those they perceive as successful or authoritative, leading to the commodification of indigenous knowledge (Bandura). Since social media has also been utilised for educational purposes, the ability to create a certain perception is influenced by the fact that it usually fails to portray the genuine image(Akram and Kumar, 2017). One prominent example is the practice of yoga for aesthetics on Instagram, where influencers frequently display difficult positions in picturesque settings without talking about the meditative and spiritual benefits of yoga. Replicating this approach, Indian influencers like @indianyogi might put engagement metrics ahead of the underlying significance of their customs, changing how viewers view and approach yoga.

Due to their frequent lack of digital literacy, traditional practitioners are unable to attract a wide audience on Instagram. As a result, more well-known Western influencers who control the conversation drown out their voices. The Spiral of Silence Theory explains this phenomenon, stating that individuals refrain from expressing views that contradict dominant media narratives for fear of isolation. A study in 2024 found that:

Integrating Indian knowledge traditions into contemporary society requires dedicated scholars, experts, educators, and consultants. The challenge lies in presenting ancient knowledge in a modern format and conducting extensive research on texts such as the Bhagavad Gita. (Singh 1791)

As a result, indigenous knowledge holders struggle to maintain authenticity in the face of widespread Westernised representations. For example, lesser-known Ayurveda practitioners in India often fail to reach a wide audience compared to Westernised influencers who repackage the same knowledge in a more digestible format, as seen with @ theguthealthmd, whose content garners significantly more engagement than traditional Indian Ayurvedic doctors.

Users witness rebranded or simplified versions of cultural practices that conform to current trends, personal opinions, as online content is increasingly created for virality and engagement. These platforms put relatability and reach ahead of cultural accuracy, which causes false customs to be widely accepted. Users are more inclined to believe and accept these narratives as accurate representations of the society they depict when they are constantly exposed to such selective content (Dennis et al.).

Authentic Indian sources are frequently less

visible and engaged on social media than the Western interpretations of Indian Indigenous Knowledge Systems (IKS). For example, despite frequently referencing ancient Indian yoga techniques, well-known Western yoga influencer @yoga_with_adriene has a sizable international following and high levels of engagement. According to De Michelis (2008), the most effective way that globalised yoga styles have advanced towards acculturation in most societies is through medicalisation. Indian online personalities like @sadhguru and @indian.yogi, who portray yoga in its traditional and spiritual context, receive less attention. The trend is also comparable in artisanal traditions. Companies like @freepeople have renamed Indian textiles and crafts as "boho-chic" to appeal to Western consumers while detaching the art form from its historical roots. Meanwhile, Indian businesses like @jaypore and @fabindiaofficial, which promote authentic, locally made goods and craftsmanship, usually struggle to get the same amount of visibility and interaction globally. This disparity shows how social media platforms usually support IKS's Centre aesthetics, displacing native voices and distorting cultural narratives. While some of the Indian brands and influencers do get recognised nationally, they fail to get reasonable viewership globally, along with their international counterparts.

Conclusion

Indian Indigenous Knowledge Systems (IKS) have gained international attention in the digital age, especially through social media sites like YouTube and Instagram. This exposure has resulted in broad westernisation, as traditional traditions are reframed, simplified, and frequently deprived of their cultural core, even while it also presents opportunities for cultural engagement and rebirth. Using media theories like the Spiral of Silence, Cultivation Theory, Framing Theory, and Social Learning Theory, this study shows how algorithm-driven visibility, aesthetic rebranding, and repeated exposure all lead to the marginalisation and misrepresentation of authentic IKS. Promoting culturally grounded narratives, elevating indigenous voices, and critically interacting with the media landscapes that influence public opinion are all crucial to maintaining the integrity of these knowledge systems. There is a genuine risk of losing the richness, diversity, and dignity ingrained in India's traditional knowledge legacy if deliberate measures are not made to preserve and accurately portray IKS. Modern tools, including Artificial Intelligence and proper algorithms, may be developed to properly monitor the trends social media propagates. The government of India has made deliberate measures to promote the education of IKS through NEP 2020 and by introducing various new courses on IKS. A proper education on the past is necessary for the forthcoming generations to identify the shifts in the portrayal of indigenous knowledge systems.

Bibliography

- Akram, Waseem, and R. Kumar. "A Study on Positive and Negative Effects of Social Media on Society." *International Journal of Computer Sciences and Engineering*, vol.. 5, no. 10, 2017, pp. 351–354. https://www.researchgate.net/publication/323903323
- Auxier, Brooke, and Monica Anderson. *Social Media Use in 2021*. Pew Research Centre, 7 Apr. 2021, https://www.pewresearch.org/internet/2021/04/07/social-media-use-in-2021/.

- Bandura, Albert. *Social Learning Theory*. Prentice Hall, 1977.
- Dennis, Alan R., Antino Kim, and Brittany Marett. *Fake News on Social Media: People Believe What They Want to Believe When It Makes No Sense at All. SSRN Electronic Journal*, Jan. 2018, https://doi.org/10.2139/ssrn.3269541.
- De Michelis, Elizabeth. "Modern Yoga: History and Forms." *Yoga in the Modern World: Contemporary Perspectives*, edited by Mark Singleton and Jean Byrne, Routledge, 2008, pp. 17–35.
- Elhajjar, Samer, and Omar S. Itani. "Examining the Impact of Social Media De-Influencing on Audiences." *Internet Research*, Feb. 2025, https://doi.org/10.1108/INTR-04-2024-0574.
- Entman, Robert M. "Framing: Toward Clarification of a Fractured Paradigm." *Journal of Communication*, vol. 43, no. 4, 1993, pp. 51–58.
- Gerbner, George, and Larry Gross. "Living with Television: The Violence Profile." *Journal of Communication*, vol. 26, no. 2, 1976, pp. 172–199.
- Goffman, Erving. *Frame Analysis: An Essay on the Organization of Experience*. Harper and Row, 1974.
- Lenard, Patti Tamara, and Peter Balint. "What Is (the Wrong of) Cultural Appropriation?" *Ethnicities*, vol. 20, no. 1, 2019, pp. 126–144, https://doi.org/10.1177/1468796819866498.
- Singh, Rajkumar. "Challenges of Indian Knowledge System in the Context of NEP 2020." *International Journal of Progressive Research in Engineering Management and Science (IJPREMS)*, vol. 4, no. 12, Dec. 2024, pp. 1789–1793, www.ijprems.com.

Representational Divergence: Shakuntala in Vyasa and Kalidasa

Dr Smithi Mohan

Associate Prof, Dept of English, Government College for
Women ,Thiruvanathapuram

The narrative of *Abhijnanasakuntalam* closely parallels earlier versions found in both Hindu and Buddhist literature. In *The Mahabharatha* the story is foundational to the lineage of the Pandavas and Kauravas, with the union of King Durhyanta and Shakuntala culminating in the birth of Bharata, the eponymous ancestor of the Bharata dynasty. Similar motifs are also observed in certain Jataka tales, suggesting a broader cultural resonance across traditions. The play's central thematic concern revolves around Dushyanta's recognition of his forgotten wife through a symbolic token, often rendered in translation as *The Lost Ring* or *The Fatal Ring* (Monier-Williams). The multiplicity of titles across editions and translations also reflects the diverse interpretive lenses through which the play has been received globally. This paper examines the contrasting portrayals of Shakuntala in Vyasa's *Mahabharata* and Kalidasa's *Abhijnanasakuntalam* through an analysis of Kalidasa's pursuit of aesthetic pleasure versus Vyasa's exploration of human suffering and moral complexity. The paper argues that these representations reflect broader cultural and philosophical shifts in how women's experiences were framed and valued in classical Indian literature.

The original story of Shakuntala is narrated in the epic *The Mahabharata*. Ved Vyasa's divided the epic Mahabhatata into 18 parvas. The story of Shakuntala can be found in the first Parva, i.e. Adi Parva or the "Book of the Beginning", under the section 'Sakuntalopakhyana'. It is the glorious story of Dushyanta and Shakuntala. The story is narrated in the ancient Indian epic *The Mahabharata*. Vishwamitra, performing intense penance, alarmed Indra, who sent Menaka to distract him. Realizing Indra's deception, Vishwamitra renounced Menaka to resume his spiritual journey. Menaka abandoned their daughter near Rishi Kanva's hermitage. In the Adi Parva of *The Mahabharata*, Kanva says: "She was surrounded in the solitude of the wilderness by Shakuntas, therefore, hath she been named by me Shakuntala (Shakunta-protected)."

The twist occurs when, King Dushyanta loses his way and reaches a hermitage where he meets the beautiful Shakuntala. She introduces herself as the adopted daughter of Sage Kanva and the biological daughter of Menaka and Vishwamitra. Dushyanta and Shakuntala marry according to the Gandharva tradition, with the promise that their son will inherit the throne. After giving birth to a son, Bharat, Shakuntala approaches Hastinapur, where Dushyanta initially denies knowing her. A divine voice (Akashvani) intervenes, compelling Dushyanta to accept Shakuntala and Bharat, thus restoring their honour.

Shakuntala's story has been adapted into various Indian art forms, including painting, dance, theatre, and film; highlighting its enduring cultural significance. Raja Ravi Varma's 1898 painting Shakuntala stands as a significant visual interpretation of the character's emotional depth. Depicting Shakuntala feigning the removal of a thorn while covertly glancing for Dushyanta, the artwork utilizes techniques such as chiaroscuro and realism to evoke longing and anticipation. Varma's synthesis of European artistic methods with Indian themes is evident in this composition. The narrative of Shakuntala has also been a central motif in Indian classical dance traditions like Bharatanatyam and Kathak. Through expressive gestures (*mudras*) and precise movements, these performances convey the emotional spectrum of Shakuntala's experience—especially themes of love, separation, and eventual reunion. Cinematic retellings of Shakuntala's story have contributed significantly to its cultural legacy. The 1943 film *Shakuntala*, directed by Shantaram Rajaram Vankudre and featuring Jayashree and Chandra Mohan, achieved both national acclaim and international recognition as the first Indian film released commercially in the United States. Later adaptations include the Telugu film *Shakuntala* (1966) starring N.T. Rama Rao and Saroja Devi, and the recent production *Shaakuntalam* (2023), directed by Gunasekhar and headlined by Samantha Ruth Prabhu.

Kalidasa (5th century ce, India) was a Sanskrit and dramatist, probably the greatest Indian writer of any epoch. Kalidasa as a dramatist gave depth to his works by making his characters multi-dimensional and expanding their scope with imaginative flights. In his writing, characters appear layered, and their actions are governed by well-conceived notions of beauty. Kalidasa adapted the story of Shakuntala from *The Mahabharata* and dramatized it as a Cinderella-esque story about a poor young girl who ends up being a queen. He has executed the story in a polished literary mould. Hence the presentation of the story bears some uniqueness. Kalidasa has divided the play *Abhijnanashakuntala*into seven acts- Act-I : The Hunt; Act-II : The Secret ; Act-III The Love Making; Act-I V Shukuntala's Departure; Act-V Shukuntala's Rejection; Act-VI Separation from Shukuntala ; Act-VIII Reunion.

The narrative in *Abhijnanashakuntalam* is in a similar vein, but Kalidasa brings in some minor but significant changes. Shakuntala lost in the thoughts of her lover, fails to notice Durvasa's arrival to the hermitage. Durvasa gets insulted and curses Shakuntala, proclaiming that the person she is thinking will forget her. Shakuntala's friends explain the situation to the sage and apologize to him for her actions. The Sage then relents and informs them that the sight of a memento could lift the curse. They are relieved as Dushyanta has gifted a ring to Shakuntala. When Sage Kanva returns, he makes arrangements to send her to her husband before Bharat's birth but unfortunately, she loses the ring when she bathes in the Ganges and consequently Dushyant fails to recognize her in the absence of the ring.

A few days later, a fisherman gets the ring from the gut of a fish, produces it before the king whose memory is restored and he laments the loss of Shakuntala and his unborn child. Dushyant discovers them at last, several years later, in a hermitage where Shakuntala has given birth to Bharat. The family gets united. Kalidasa unlike Vyasa has meticulously filled the gap in *The Mahabharata* by explaining the reason for Dushyanta's conduct. He has exonerated Dushyant by explaining his irresponsible behaviour and softens Shakuntala's character. The ring plays a crucial role in exonerating Dushyanta. It is the signet ring Dushyanta gave to Shakuntala as a token of their love, which she loses in a river. Later, a fisherman finds the ring in a fish, and it is brought to the king, who upon seeing it, remembers

Shakuntala and is freed from a curse that made him forget her. Shakuntala is presented as a shy character throughout the play – she is portrayed as an innocent, gentle girl, living in harmony with nature. This is in total contrast to her self-confident and fiery portrayal in *The Mahabharata*.

Kalidasa has borrowed the theme from *The Mahabharata* and made some interesting changes to the original to give the play a dramatic look, and provide a justification for Dushyanta's behaviour. But the main focus of this paper is the etching out of Shakuntala's characterization which is very different from the earlier heroines of Sanskrit literature. The Shakuntala of *The Mahabharata* is not like Cindrella, for she belongs more to the twenty-first century than the fifth century or can even be regarded as a timeless character. *The Mahabharata* does capture the cold and heartless treatment meted out by Dushyanta to his wife and son. There is no explanation whatsoever for the same. Perhaps, the only reason would be that Dushyanta being a human is flawed.

Romila Thapar reiterates this and says that in *Abhijnanasakuntalam*, not only does the context and the story change but, more pertinently, the character of Shakuntala is a contrast to the woman portrayed in the epic. There is a resistance as if to the epic version through the presentation of modern times and the Shakuntala of the epic is seen to be marginalized. As Thapar writes, "in Kalidas's version we are in the realm of delicacy and romance, imminent tragedy and finally happiness" (43). The emotional range is infinite when compared to the epic narrative, but while intermeshing of the emotions, we can see the image of Shakuntala undergoes a transformation. In the play, Kalidasa's interest lies mainly in portraying the budding development of love between Dushyanta, a sophisticated and noble king and an innocent girl. The journey moves from the hermitage and the bond between the duo is strengthened through separation and suffering.

The Mahabharata's Shakuntala is a woman in a patriarchal society as is Kalidasa's Shakuntala. But in *The Mahabharata* she seems to have carefully thought over the roles of wife / mother and son in a society dominated by men such as Dushyanta during the six years, when she lived forgotten by her husband. When she goes to Dushyanta's court, she is the mother of a six year old boy but her behaviour at the court shows that she has not lost her strength or hopes for her son. But also her words to Dushyanta are no longer those of a young woman innocently and happily wandering through her father's hermitage. After long separation she tells Dushyanta about their son: "This is your son, O king, he should be consecrated by you as your heir".

On the other hand in Kalidasa's play, Shakuntala after the cruel rejection by the king, is stunned, filled with shame and sorrow, followed by anger. There appears to be highlighting of her patience and her verbal skill in her characterization. Then she quickly conceals this anger, and gains control over herself. Her speech is long and impassioned, but rational. She is not whining or begging the king. She boldly tells Dushyanta of the future implications and results of his false action. "O great king, even though you do recognize me, why do you say: "I do not know you?" You speak thus carelessly as another, a low-born villain might speak". She further, says, "if you will not follow my advice O Dushyanta, you will reap the results of your present actions a hundred times". Shakuntala's strength echoes in her words at the end of the story. Divine intervention allows one to believe that God respects and takes due cognizance and care of her words. Shakuntala's skills of argument and of drawing upon associations of memory can be seen in the play when she is in the court, is pregnant and the king refuses to recognise or accept her. She is stunned but does not say a word. But the hermits and Gautami tell the king that a virtuous woman is believed to be evil by the world if she continues to live with her own family. She also first tries to show the ring but discovers it is lost. Then reminds him of the simple forest scene with the buck (fawn). She continues to tell him that she is his wife. Dushyanta at the end of the court scene, does not accept her. But she, through her argument, verbal skills and emotional appeal puts him in a painful and pensive mood. Dushyant says "I cannot remember marrying the sage's abandoned daughter, but the pain my heart feels makes me suspect that I did."(45)

In Kalidasa's play, Dushyanta blames her for her untrustworthy character and her parentage, Dushyanta insultingly remarks to Shakuntala that women are born cunning. This unrighteous action of Dushyanta draws out an angry response and one can see Shakuntala's head held high, her eyes flashing when she answers the king. "You point out the faults of others, even though they are as small as mustard seeds, but you do not notice your own faults, which are as big as bilwa fruits". She further says, "my mother Menka is a celestial. My birth, therefore, Dushyanta, is far higher than yours. Your place is earth, but mine is in the sky." Shakuntala, after Dushyanta's rejection warm-heartedly tells him "I was in infancy cast away by my parents, and now I am cast away by you! Well I am ready now to return to the ashram of my father, but you must not cast away this child who is your own."(27)

Kalidasa's Shakuntala is different from earlier

heroines of Sanskrit literature. Even though Dushyanta does not recognize her, her reaction is different. She is not angry and does not shout at the king but says she caused her situation herself that she trusted the king and starts weeping. She does not beg for pity from the king but weeps and cries out: "O mother earth, give me room (in your bosom)" These words echo those of Sita of the Ramayana rather than Vyasa's Shakuntala but in Kalidasa, Shakuntala is whisked away by her apsara mother Menaka to heaven. Her caring and nurturing quality, her key strengths are beautifully described in the opening scene of the play when she is watering the plants / flowers in the hermitage and that is why Kanva at the time of departure to palace address the trees accordingly; "This Shakuntala, who ever wished to drink water while you are all yet un-watered; who fond though she is of decorating, never plucks a sprout from you through affection, who enjoys festal celebration on the occasion of your first blossoming she, this Shakuntala, is today leaving for her husband's palace. Pray, let her receive the permission of you all. (15)

In *The Mahabharata*, Shakuntala's boldness is seen at Dusyanta's court when Dushyanta does not accept his son, terms of duty pleasure or profit saying Shakuntala is lying: "Women don't tell the truth. Who will take your word for it?" He casts doubt upon the story of her birth and says she is rather of low birth, an evil ascetic." Shakuntala tells him again and again that she is well born and he is being obtuse and finally she boldly answers, "This broad four-edged earth crested by regal crags, will be governed by my son whether you like it or not, Dushyanta!" After this a disembodied voice in the sky announces that the son is Dushyanta's and that he must accept him and Shakuntala. In Vyas's Shakuntala, she is reunited with her husband in the palace itself. But in Kalidasa's play she breaks down and in her anger and distress calls for her mother earth to give space in her bosom and finally her mother Menaka protects her with "a flash of light in a woman's shape".

The primary goal of art, according to Rasa theory (Bharata, 9), is to evoke particular emotional responses in the audience, creating a shared emotional experience between the creator and the viewer. According to rasa theory, the emotional essence of a play is conveyed through the audience's response to the characters' emotions, which can be cultivated through aesthetic means such as facial expressions, gestures, and dialogues. The protagonist of Kalidasa's Shakuntala is portrayed as an idealized character whose role as a soon-to-be mother is crucial, rather than just as a woman with agency. The story is infused with poetic beauty from the outset, making Shakuntala a symbol of maternal virtue and purity.

Shringara rasa (the sentiment of love and beauty) permeates the earlier sections of the play, especially in Shakuntala's blossoming romance with Dushyanta. Their mutual attraction, filled with coy glances, tender exchanges, and moments of longing, embodies the idealized vision of romantic love.Adbhuta rasa (wonder) is evoked through the portrayal of divine interventions, lush forest settings, and Shakuntala's celestial parentage. Shakuntala's pregnancy is not merely a private experience; rather, it is entwined with the rasa of karuna (pathos), evoking a sense of sympathy and sorrow that amplifies her role as a mother-to-be.

The rasa of karuna is most prominent in her plea to the earth when she calls out: "O mother earth, give me room (in your bosom)." These words evoke a sense of helplessness and dispossession, echoing Sita's own lament in the Ramayana. The phrase expresses a connection to nature and the divine as a form of refuge, echoing Sita's similar cry when she experiences disempowerment and abandonment as well. However, because Shakuntala's maternal identity is encased in an idealized, aestheticized framework, the echo of Sita's words is even more noticeable in Kalidasa's text. Here, the rasa is designed to emphasize the poetic grandeur of her reproductive role in addition to providing emotional engagement.

However, the focus on Shakuntala's motherhood is less obviously aestheticized in Vyasa's Shakuntala. Her pregnancy is portrayed as the consequence of her lover's abandonment and the ensuing trauma she experiences, making her story more rooted in the difficulties she encounters. In contrast to Kalidasa's Shakuntala, Vyasa's Shakuntala does not scream out. The tone is more melancholic and less focused on beauty, which is more in line with Vyasa's realistic depiction of women's suffering under patriarchal social norms. Kalidasa succeeds in creating an emotional connection between Shakuntala and the audience. By invoking specific rasas such as karuna (pathos) and adbhuta (wonder), Kalidasa manipulates the emotional flow of the narrative to emphasize Shakuntala's maternal purity. Through the aesthetic tools of poetic language and visual representation (such as the lush forests, the celestial setting, and her divine interactions), Shakuntala is framed as an object of veneration, rather than as an agentic subject.

Kalidasa neutralizes Shakuntala's agency and reduces her complexity to a single role within the greater story of royal lineage by emphasizing her symbolic role as a soon-to-be mother. This aestheticization, while beautiful and emotionally resonant, limits her capacity for self-determination. On the other hand, Vyasa's depiction of

Shakuntala provides a more realistic and insightful analysis of a woman's experience of abandonment and how it affects her identity. Vyasa's Shakuntala is more than just a mother; she is a woman grappling with loss, with the realization of her own vulnerability, and with a profound sense of injustice. Vyasa's Shakuntala is not merely an emblem of lineage but a woman who questions her fate, voices her pain, and reveals the vulnerability and resilience inherent in her experience. One cannot but allude one among the possible reasons of this divergence to be the difference in narrative purposes—Kalidasa's work serving courtly aesthetic ideals, while Vyasa's epic explores moral complexity and human suffering—reflecting broader cultural and philosophical shifts in how women's experiences were framed and valued.

List of Works Cited

- Bharata. *Nātyaśāstra*. Translated by Manomohan Ghosh, Asiatic Society of Bengal, 1951.
- Kalidasa. *Abhijnanasakuntalam* (The Recognition of Shakuntala). Translated by Arthur W. Ryder, Harvard University Press, 1912.
- Miller, Barbara Stoler. *Theatre of Memory: The Plays of Kalidasa*. Columbia University Press, 1984.
- Narayana Rao, Velcheru, et al. *Classical Telugu Poetry: An Anthology*. University of California Press, 2002.
- Pollock, Sheldon. *The Language of the Gods in the World of Men: Sanskrit, Culture, and Power in Premodern India*. University of California Press, 2006.
- Thapar, Romila. *Shakuntala: Texts, Readings, Histories*. Women Unlimited, 2010.
- Vyas, Krishna-Dwaipayana. *The The Mahabharata*. Translated by Kisari Mohan Ganguli, Munshiram Manoharlal Publishers, 1986.

Similarities between Women and Nature: An Environmental Perspective

Dr. Sreelarani M.S.

Assistant Professor, Department of Malayalam,
VTMNSS College, Dhanuvachapuram
sonianil06@gmail.com

Folklore is a comprehensive expression of the lives of ordinary rural people. It is traditionally inherited and orally transmitted. The local knowledge and literature of the common people are more extensive and richer than those of the educated. Its essence is innocence. Folklore is a mirror of the rural mind, reflecting the life of the community. Therefore, that community is the life of folklore.It reflects the culture that sees women as nature. Folklore has the ability to answer all kinds of environmental problems. One issue that needs serious thought in that regard is the struggle for the existence of women and nature. The manifestation of such strong struggles can be seen in every aspect of folk literature and art.

Let's take riddles as a special type of folklore. If you look at any of the riddles like 'Kaattiloramma Ponnaninjunilkkunnu', 'Amma Kallilum Mullilum Molu Kalyanappanthalil', 'Kattilorurvashi Kannekhitippottumthottu', we can see that all of them are characterized by the comparison of women and nature. Apart from this, this comparison is evident in the descriptions in the story and the poem. Such descriptions are more common in songs, especially in Northern ballad songs. The active presence of nature can be seen wherever women are described like "Kunnath Konnayum Pootha Pole', 'Poopolazhakullaval', 'Kunnikuruvin Varnam Pole', 'WayanadanManjal Muricha Pole'. The beauty of a woman's body is compared to every object in nature. The konna flower in the forest, the turmeric in the soil, and the coconut flower clusters are all presented as symbols of the prosperity and abundance of womannature. These appear as gentle expressions of femininity. However, the fierce nature of a woman is revealed in the expressions used for cooking rice, Chempil Bhagavathi Thullithulli,' and for the fire burning in the hearth, ' Karinkaalikku Moonnumulayum Theenaavum.' These are descriptions related to the kitchen.

The distinction between the inside and outside of the house represents nature and culture. The contradiction between inside and outside was later created as part of the idea of culture. However, in a time without systems or controls, women were more independent. In the patriarchal social system, the level of women's subjugation and aggression towards them increased, and it disrupted the balance of nature itself. However, one important thing to understand here is that the centre of life is the family, and the centre of the family is the woman. In such a centralized situation, all the economic and environmental problems of the family are related to the woman.

The biggest challenges faced today by women and nature, who play a major role in the stages of production and growth, are exploitation and pollution. Womanhood, which toils for the reconstruction and restoration of society, is being marginalized from the mainstream of society. For women, production and reproduction are not just physical activities, but also social efforts for survival. Creation and protection are equally important for women and nature. Recognizing the problems of humanity points to the need for the protection of women and nature. Because women and nature are the protectors of life on earth. Still, the question of why they are both so exploited is still relevant today. In this era, when there are more efforts to create awareness about the protection of nature and women, they are being exploited more. The reason for this is that both are experienced by men as mere physical bodies. The idea that nature was formed for the convenience of his material life and woman was formed for the satisfaction of his bodily desires should be eradicated first. Women empowerment is relevant only in an era where there is awareness that there is no world without nature and women.

By presenting each aspect of nature with femininity, the greatness of the agricultural culture that sees woman as a symbol of fertility and prosperity is revealed. The worship of the earth and natural phenomena is rooted in ancient mother worship. This culture is also evident in fertility worship. Here, nature creates a sublime world and protects women, while simultaneously creating an otherworldly realm where women are revered. The woman who is the producer and protector of life has to protect nature. In this way, woman and nature function as mutual complements. Exploitation of those who fulfil human needs is tantamount to self-destruction. Man should treat nature like children with love and a sense of duty towards their loving mother. When the opposite happens, she becomes destructive. This is nature's reaction, a reaction when things go wrong.

A community that worshipped mother goddesses and nature deities for happiness and to be saved from diseases disrupted the protection of women and nature, leading to many environmental problems. Here, the erosion of values of a culture that had lasted for ages was beginning. Man's eagerness to establish dominance over women and nature became the cause of creative destruction. Those who seek the cause of natural disasters like floods, landslides, and droughts would do well to study and evaluate the exploitation of nature. It is a serious offense to encroach upon nature and the habitats of living beings and to obstruct their free movement. News of wild elephants entering villages and destroying crops, and packs of wild boars attacking people are often seen in newspapers and other media. It can be understood that this happens to a certain extent because humans interfere with their habitats. Another reason is that modern humans, who boast of being city dwellers, have started to interfere with the habitats of villagers who are degraded as tribal. They lived in accordance with nature. Modern man carried out his exploitation in the form of invasion there too. They rejected their beliefs. The lifestyle was completely destroyed. What happened to the well-organized ecosystem? Due to overexploitation, nature began to change as it pleased. Nature started to trouble man with summer when DAP was due and rain during summer. The balance of nature was completely upset. But the consequences had to be faced by the helpless and destitute people. The lessons given by Churalmala and Mundakkai in Wayanad are unique. Environmental activists who studied and gained knowledge and awareness about the Western Ghats were not taken seriously by anyone. Environmental activists could only worry about its protection. Eventually, everyone realized the meaning of what the environmental activists said when the lives of some people were lost.

The case of women is not different; on one hand, women are praised as mothers and gods, while on the other hand, they are exploited. In the contemporary cultural landscape of Kerala, the life of a woman is more difficult than before. Pitfalls lie everywhere. The very hands that should protect her become her noose. She has transformed into Kuriyedathu Thathri, a vengeful goddess who confronts the man who comes to exploit her. The law has come to her aid to bring the deceit of men before society. Yet, he exploits her from beyond the bounds of the law. A woman, simultaneously strong and weak, and enslaved by emotions like affection, did not require much effort from a man to be subdued. However, she rose up, reminiscent of the tale of the Phoenix bird. Time elevated her to a personality unwilling to be defeated again. Remembering the women who stumbled and fell along the way, the modern woman built her own space. There, she began to live for herself. She transformed from the silent pain of yesterdays to the firm voice of today. Nature, exhausted from enduring all the blows inflicted by humanity, also took on its own protection.

Women and nature are not two separate entities and unless men feel that it is their responsibility to protect them, the balance of the earth will remain uncertain. Environmental protection is never possible in a society where misogyny prevails. Interdependence and cooperation bind men and women.

Only when the understanding that women, men, and nature are protectors of the universe arises, will a balanced mindset be born. As long as that doesn't happen, the fact that women and nature are changing in a way that challenges the patriarchy that exploits them is emerging as a serious environmental problem.

Reference Books:
- Folklore : Dr. Raghavan Payyanad, Kerala Bhasha Institute, Thiruvananthapuram.
- Folklore Samskaram: P Soman, Kerala Bhasha Institute, Thiruvananthapuram.
- Vaakkum Naattarivum : Utharamkodu Sasi, Kaliveena Publications.
- Folklore Puthuvazhikal Navavaayanakal (ed.): Jobin Chamakala, Turn Books, Kottayam

Black Eagle Books

www.blackeaglebooks.org
info@blackeaglebooks.org

Black Eagle Books, an independent publisher, was founded
as a nonprofit organization in April, 2019. It is our mission
to connect and engage the Indian diaspora and the world at
large with the best of works of world literature published
on a collaborative platform, with special emphasis on
foregrounding Contemporary Classics and New Writing.